Keeping
Your
Eye
on
Television

Keeping Your Eye on Television

by Les Brown

The Pilgrim Press · New York City

Library of Congress Cataloging in Publication Data

Brown, Les, 1928–
 Keeping your eye on television.

 1. Television broadcasting—Social aspects—United
States. 2. Television broadcasting policy—United
States. 3. Television programs, Public service.
I. Title.
HE8700.8.B748   384.55'4'0973   79-15828
ISBN 0-8298-0376-9

The Pilgrim Press, 132 West 31 Street, New York, N.Y. 10001

To the memory of Bill Greeley

Researched by Arthur Greenwald

Contents

Introduction

It is generally conceded that broadcasters are a powerful force in any community. They have more influence on values than does any other institution, including the church. They control the flow of information and opinion to more people than does any other medium of communication. They arouse our desires, direct our tastes, and convince us of needs we never imagined we had. They make and break politicians and public figures. They are eagerly wooed by business to sell products and corporate images.

A television license is perhaps the most valued gift our nation bestows on a citizen, so valuable that even incredibly incompetent management cannot avoid making large yearly profits.

Broadcasting was founded on the premise that the holders of station licenses, in return for the free use of broadcast frequencies, would operate "in the public interest, convenience and necessity." The people would benefit by having a forum for debating controversial public issues and a medium for cultural enrichment. What we got was a money machine.

It would be ludicrous to argue that television should be devoted solely to the uplift of society. Nor should we underestimate the value of good entertainment as a positive social force. Commercial television provides such good entertainment in shows as *All in the Family, Paper Chase, The Waltons, Police Story,* and *Edward VII.* But such programs are only a small component in what Newton Minow, a former Federal Communications Commission chairman, once dubbed "a vast wasteland."

Unfortunately, in its effort to squeeze out maximum profits, television has largely turned its back on its public interest responsibilities, and the FCC has allowed it to do so.

The Office of Communication of the United Church of Christ has worked for more than a decade to foster mutually beneficial exchanges between broadcasters and public bodies about needs, tastes, and desires in programming. More than one thousand community leaders in all parts of the country have participated in "Check Your Local Stations" workshops under Office of Communication sponsorship. There they have received training about government regulation of broadcasting, how to evaluate local station performance, and how to organize their communities to work for improved television services and for jobs in stations for minorities and women. Licensees are now being visited regularly by knowledgeable civic organizations that want to discuss with station officers how to deal with community needs. These public bodies are determined to have their views considered with respect, and to bring about change in broadcast practices.

One of the most important issues that is now before Congress is reform of the Federal Communications Act. There is general agreement that a communications act formulated in 1934 is inadequate to deal with today's technologies, with the growing competition between telecommunications systems (such as cable TV and over-the-air TV broadcasting), and with future problems that will grow out of the two.

It remains to be seen whether revisions of the act will embody the interests of consumers, or whether they will be wholly to the advantage of broadcasters, cable operators, telephone and satellite companies, and other users of the electronic spectrum.

Only those persons who will address themselves to the questions that are ultimately essential and who will clearly define the issues will influence the making of telecommunications policy. Among the essential questions are:

> Is deregulation the remedy for poorly administered regulation?
> Can we legislate effective competition?
> Do we foster diversity of opinion and experience simply by proliferating new technologies for distributing messages?
> Can we afford to discard the basic principles of regulation that were set down in 1934?

Decisions on such questions will be made over the next few years. Citizens who are prepared can play an important role in influencing them.

Keeping Your Eye on Television is a guidebook to television for public-spirited citizens who want to participate in decisions that affect their lives. Its objective is to help them wrestle with the vital social issues that center in how this country regulates electronic communication.

EVERETT C. PARKER

Remember that in the late 1940's the Commission made most of its long term policies on commercial television. Twenty years later, the general public became concerned and aware but too late to participate. Do not forget this lesson.

Charles D. Ferris, Chairman
Federal Communications Commission,
in a speech to the ACT Symposium
in Washington, D.C., May 1, 1978

Is the Public Interested in the Public Interest?

A short news article in 1953 reported that a man with no criminal record went for his gun and shot his TV set. It was the first known instance of videocide, although not the last, and it bespoke the frustration of viewers around the country—in those days when television was young— who sat before their sets unable to resist the talking images but helpless to answer back when those images and what they were saying (and selling) became infuriating.

Undiminished after a quarter century, that general frustration is in many households now mixed with anxieties about what television may be doing to children, the political system, or the moral fabric of society. Some writers, cursing television as a drug, have advised discarding the television set, as if that would solve all the problems. The idea is as myopic and unrealistic as the notion held by many people that they can make their statement against television simply by not watching it.

In the first place, the relatively few homes that would banish TV would not in any way diminish the medium's impact on themselves, other people, or the major institutions of society. Politics, government, business, education, religion, culture, and sports have all been changed by television and continue to be shaped by it. Jimmy Carter, it is generally conceded, could not have been elected President without television. Those who unplugged their television sets were not exempted from his presidency.

If a million households—or even five million—were to disconnect their television receivers there would still be television, and the medium would continue to affect their world and influence their lives and those of their children, grandchildren, and great-grandchildren. Moreover, no meaningful statement is made by tuning out, because that is a form of inaction, not an action. No pressure will yield no result.

In the second place, the issue in the modern world should not be how to make television go away but rather how to live with it so that it increases, not diminishes, life. People who focus their anger on the technology misapprehend the problem. The technology of television has of course dramatically changed the way we live, but if it is to be resented for that so also must the technologies of the telephone, the automobile, the jet airplane, and the computer be resented. What makes television the "boob tube" to its detractors is not electronics or the size of the screen; it is that those privileged to use the extraordinarily powerful communications medium have, for reasons of their own venality, chosen to let it carry very small freight.

By now it is clear that the television industry will never by itself realize the medium's promise of operating in the best interests of the entire public and contributing to the cultural, intellectual, moral, and spiritual enrichment of the society. Almost from the beginning, commercial broadcasters have winked at this promise and concentrated on harnessing television to exploit their markets. The medium is being milked now for more than $6 billion a year.

Whether or not we watch television, we are all, one way or another, in the medium's thrall and subject to its commercial imperatives. Television is not a mechanical appliance; it is not a toaster or a Waring blender with a picture. It is an environment. What happens on television is as real, in its implications and its influence on everyday life, as weather, the quality of air, or a plague of the gypsy moth.

Violence becomes part of the social climate if violence appeals to the segment of the television audience advertisers are most eager to reach—the eighteen- to forty-nine-year-old group. If violence represents an opportunity to maximize audience and thereby maximize profits, then violence is churned out endlessly on home screens with little concern for its possible effects on the heaviest users of television—children and the elderly—or on society in general. If sexual titillation supplants violence as the flavoring most seductive to viewing, then it is sex—expressed in innuendo, body movement, and revealing dress —that becomes the medium's preoccupation.

But the broadcasting system in the United States was not conceived by Congress as one that would strive primarily to please the eighteen- to forty-nine-year-old consumer. Nor was it intended that radio and television operators be allowed to pursue their own commercial interests without regard for the needs of the communities they are licensed to serve, nor that all owners, managers, announcers, and newscasters be male and white, nor that there be no accountability to the public.

The fact that thousands of letters of complaint are received by the Federal Communications Commission (FCC) each month suggests that some people recognize, however vaguely, that they have some rights in the media of radio and television and that the licensing of broadcast stations may have something to do with accountability.

The people do have rights, because the airwaves over which radio and television stations broadcast belong to everyone. It is the same air as the air we breathe, and no one with a license may claim it as his or her private property. Broadcasters are given the privilege of using those airwaves, with their limited number of frequencies, as public trustees. It immediately becomes their responsibility under the law to use the air in ways that serve "the public interest, convenience and necessity." The Federal Communications Commission was created to manage traffic on the electromagnetic spectrum, to award licenses to the

7

most qualified applicants, and to see that the public interest is served.

A 1963 House of Representatives report on the FCC stated:

> Under our system, the interests of the public are dominant. The commercial needs of licensed broadcasters and advertisers must be integrated into those of the public. Hence, individual citizens and the communities they compose owe a duty to themselves and their peers to take an active part in the scope and quality of the television service which stations and networks provide and which, undoubtedly, has a vast impact on their lives and the lives of their children. Nor need the public feel that in taking a hand in broadcasting they are unduly interfering in the private business affairs of others. On the contrary, their interest in programming is direct and their responsibilities important. They are the owners of the channels of television—indeed of all broadcasting.

Since that is the case, it is appropriate to ask why broadcasters generally have not honored their pledge or kept the promises made when they received their licenses. The answer, admittedly oversimplified for the moment, is that the system envisioned for American broadcasting in the Communications Act of 1934 has never been made to work.

Over the years, the FCC, whose members are political appointees, has by turns been negligent, indolent, and highly susceptible to the lobbying of the broadcast industry. Citing ratings that show meager viewing for serious programming, broadcasters have convinced the regulators, as well as themselves, that the public is not interested in the public interest.

Indeed, by its failure to assert itself over the years, the public has seemed not to be interested. The passivity may be attributed partly to people's ignorance of their rights in broadcasting but partly also to their misplaced trust in the federal agency to regulate conscientiously and competently in their behalf. Meanwhile, the press, as the public's representative, kept its eye on the TV screen and not on what was going on behind it. Had it given the institution of television the same careful scrutiny it has given the institution of government, the press might have caused broadcasting to develop differently.

Minimally regulated in the public interest, commercial radio and television, which were intended to provide vital services to their localities, have declared themselves advertising media. American broadcasting has thus developed as a hugely profitable and powerful government-protected oligopoly.

Because the regulatory system has failed, some policymakers in Washington have been prepared to write it off as a failure. In 1978 the House Subcommittee on Communications considered a bill to replace the 1934 Communications Act with a new one that would give broadcasters permanent licenses (instead of three-year licenses subject to renewal) and relieve them of their obligations to perform in the public interest. In exchange for this freedom, the bill proposed to charge broadcasters annual fees for their use of the electromagnetic spectrum. Those fees were to symbolize a payment to the public for the use of their airwaves and, under the proposal, were to have gone into a fund that would be used, among other things, for financing public television programming and assisting minorities in purchasing radio and television properties. The fact that the television industry was not wildly excited about the plan, and in fact opposed it, indicates how comfortable television broadcasters are with the existing regulatory scheme.

Citizens-action groups such as the Office of Communication of the United Church of Christ denounced the bill as a giveaway of the public's rights. The charge was minimized by a key figure in telecommunications policy, who remarked that it could not be much of a giveaway "since most of the public was never really aware that it had rights." But a sufficient number of private citizens have become aware of their rights, and their involvement in broadcast affairs since the late 1960s has made a substantial difference in the responsiveness of the broadcaster to the needs of the society. Working through the system, citizens groups have begun to make the system work.

Through the efforts of these groups, women and ethnic minorities have begun to assume larger roles in the media, both on the air and in management ranks. Viewpoints contrasting with those of the station licensee or the local power establishment have been given voice. Networks and stations have instituted reforms in children's advertising and programming. Programs and old movies offensive to ethnic groups and homosexuals, fueling the prejudices that have caused them suffering, have been removed from the air or revised. Violence on television has lessened. Overall, there is a healthy interaction between licensees and their constituencies—or at least the beginning of one—that did not exist before the 1970s and that has given meaning to the public-interest standard.

All this has happened because some citizens recognized that by asserting their statutory rights they could become an effective force in the broadcast system, a conscience imposed on an industry obsessed with increasing its revenues and profits. These activities have had the endorsement and encouragement of the courts.

Several years before he became chief justice of the United States, Judge Warren Burger wrote in a 1966 opinion of the U.S. Court of Appeals for the District of Columbia:

> Such community organizations as civic associations, professional societies, unions, churches, and educational institutions or associations might well be helpful to the Commission. These groups are found in every community; they usually concern themselves with a wide range of community problems and tend to be representatives of broad as distinguished from narrow interests, public as distinguished from private or commercial interests.

The job calls for reasonableness and an understanding of where citizens' rights end and censorship begins. It also calls for a watch on television and not just a watching of it. Finally, it helps immeasurably to know how broadcasting functions as a business, what federal regulation is all about, and where the citizens-action group fits into the picture.

2

The
Public
as
Merchandise

Through most of the history of American broadcasting, the people's voice in the media has been the audience ratings. People write letters to stations, networks, and the FCC, but nearly always to no avail. The only feedback from consumers heeded by operators of radio and television stations is the answer to the question How popular are we? Qualitative views and complaints are largely irrelevant. What counts—because it puts money in the purse—is how many people are watching or listening to a broadcast.

Ratings are estimates of audience size that are projected from representative samples of the populace. At one time this research was conducted by telephone; now it is done electronically by the A.C. Nielsen Company and Arbitron, with meters attached to television sets sending their information to computer centers. These results are collated with data gathered from separate surveys of families who agree to keep diaries of the daily viewing of each member of the household. This makes it possible to estimate not only the total number of people watching but also the number of viewers according to age, sex, and education and income levels.

Because they represent viewers, ratings are what the advertiser buys. Rarely does an advertiser today purchase —that is, sponsor—an actual program; indeed, many advertisers may not even know what programs their spots appear in. The transactions between the broadcaster and advertiser are conducted on the basis of anticipated audience figures.

The advertising rates for a commercial are usually governed by the "quality" of the audience it reaches. An audience predominantly of young adults in the eighteen-to-thirty-four age bracket commands the highest rates, because that group is the prime target of most of the kitchen, cosmetic, automotive, and drug products advertised on television and radio. Next in value is the larger demographic group spanning ages eighteen to forty-nine. Persons in this age range are presumed to have young families and to be stocking their households and garages with the paraphernalia of middle-class life.

Advertising rates drop substantially for audiences largely composed of juveniles or persons over fifty. These patterns explain why children's programs tend to be consigned to a Saturday or Sunday morning ghetto (prime time being too valuable for such low-yield fare) and why programs that might have special appeal or value to senior citizens are practically nonexistent on network or local station schedules.

A certain cynicism in the broadcast industry grows out of these marketplace verities. Those who conduct the business perceive the audience not as consumers of radio and television but as the merchandise of those media, a mass to be packaged and sold at rates ranging from four to twelve dollars a thousand. Getting the largest audience, and by whatever means the greatest number of young adults, has become the game.

Programs lose any purpose except that of luring viewers. Moreover, in the frenzied competition of network television, every program aspires to destroy two others that oppose it on the rival networks. TV series are not artistic creations in the sense that books and plays are, but rather have become forms of weaponry employed in complex strategies for the benefit of the advertising industry and the investment community. Movies and popular songs may be creatures of commerce, but they enjoy the independent existence of artistic works. Television programs, alone among the popular arts, are born to engage in lethal combat.

As national program services, the networks—ABC, CBS, and NBC—are the motor that drives American television, although the system was founded on the idea of localism. More than 720 television stations have been licensed expressly to provide service to their local communities, and their responsibilities in the public interest are clearly stated as nondelegable. The stations are in theory responsible for everything they put on the air, regardless of its source. Yet the vast majority of the stations give up more than 60 percent of their air time to the networks, and most of the unaffiliated stations concentrate on playing reruns of programs that have completed their network runs.

As for local service, the typical station devotes less than two hours a day to local broadcasts, most of which take the form of newscasts.

Stations yield their time to the networks because it is good business to do so. Network programs, produced on large budgets and widely promoted, attract huge audiences and bring glamour and prestige to the stations. Besides, the networks pay the affiliated stations for the air time with a share of their advertising revenues, and they provide open minutes in and around their programs for the stations to sell locally. Because the audience is bound to be sizable, those local spots sell at premium rates. The stations thus are able to make tidy profits—without risking capital on programs of their own—simply by pressing a button that brings in the network.

It is a neat business arrangement that appears to benefit everyone. But it flouts the principle of localism and also the entire concept of a public trusteeship, for it presents a situation in which an unlicensed entity—the network—decides what will be presented most of the day on licensed media. If a network should determine that it must, for competitive reasons, broaden its standards of program acceptance and deal more liberally with sexuality, two hundred stations automatically embrace the decision. The network standard becomes theirs, a case of the tail wagging the dog. In broadcasting, to stand on principle and resist the network carries economic penalties.

The networks are not licensed by the FCC because they have no direct access to the airwaves except through stations. Nevertheless, they are vulnerable to regulations through the five television stations each one owns. For example, when the FCC elected to punish CBS for intentionally misleading the public with a series of special tennis matches erroneously billed as "winner take all" events, it applied the sanction to one of the large and powerful stations owned by the network, KNXT, Los Angeles. The station received a one-year license renewal, instead of the normal three-year renewal, and a demerit. All other CBS-owned and affiliated stations, which put the tennis matches on the air and profited from them, were excused from the network's sin.

To be accurate, the networks do not actually own stations; their parent companies do—CBS Inc., Radio Corporation of America, and the American Broadcasting Companies, Inc. The networks and the station groups are separately managed sister divisions of those conglomerates, whose other businesses extend to publishing, phonograph records, toys, amusement parks, and manufacturing.

As major profit sources of these giant companies, the networks and their related stations owe their first obligations to the stockholders, who are at the top of the business pyramid. Second priority goes to the source of the money: the advertisers. Third priority goes to the FCC for the preservation of licenses. Fourth to be served—unless a part of the business equation has been overlooked—is the public.

That is how the business works in reality, but it is not how the system was intended to work.

The American System of Broadcasting

The first broadcasters in America were hobbyists and small-time entrepreneurs who needed nothing more to go on the air than the technical know-how and a little money to build transmitters. Their notable contribution in the 1920s was chaos on the airwaves, the result of broadcasting with unrestricted power and often on the same frequencies, where the clash of signals rendered their transmissions unintelligible. Calling for extensive controls, Herbert Hoover, who was Secretary of Commerce at the time, observed, "This is one of the few instances where the country is unanimous in its desire for more regulation."

Congress responded by creating the Federal Radio Commission (FRC) in 1927 to direct broadcast traffic. When there was clear need for an all-embracing national policy for broadcasting, Congress enacted the Communications Act of 1934, which among other things replaced the five-member FRC with a more powerful, seven-member agency: the Federal Communications Commission. Given authority over all wire and wireless transmissions—telephone, telegraph, and radio—the FCC was made responsible for setting the technical standards for those industries and supervising their orderly growth.

Recognizing, at the same time, that the electromagnetic spectrum used by radio was a limited resource, the Communications Act required those who received the privilege of using the public airwaves to operate "in the public interest, convenience and necessity." That phrase, borrowed from legislation regulating the railroads and public utilities, goes to the essence of the American policy for broadcasting.

The phrase is abstruse, but the Congress had expected it to gain substance and clarity from subsequent rulings by the FCC. The clearer definition of the "public interest" concept that has emerged over the years was shaped as much by the courts as by the commission. It is a definition with three premises:

- That a fundamental purpose of broadcasting is to foster informed public opinion through the dissemination of news and opinion concerning events and issues of the day

- That in order for the public to be fully informed, the stations must offer the various contrasting viewpoints reasonable opportunities to be heard

- That broadcasting is primarily a local institution, and that broadcasters, as public trustees, must work at determining the needs of their communities and program to serve those needs

These notions have been challenged by broadcasters, especially when they have been expressed in the form of regulations. The argument against them is that they limit a broadcaster's First Amendment rights of free expression. The complaint has merit, but the "public interest" standard rests upon the paradox that the rights of the few broadcasters must be abridged in order to protect the free expression of the many in the listening and viewing public. It is a controversy that continues without end.

The public-interest requirements, and the FCC's right to impose and enforce them, have repeatedly been upheld by the courts. In the landmark 1943 decision that forced

NBC to divest itself of its "Blue Network," the Supreme Court affirmed the desirability of diversity and the FCC's obligation to determine how licensing may be dedicated to the public interest. Justice Felix Frankfurter wrote for the majority:

> We are asked to regard the Commission as a kind of traffic officer, policing the wavelengths to prevent stations from interfering with each other. But the [Communications] Act does not restrict the Commission merely to supervision of the traffic. It puts upon the Commission the burden of determining the composition of that traffic. The facilities of radio are not large enough to accommodate all who wish to use them. Methods must be devised for choosing among the many who apply. And since Congress itself could not do this, it committed the task to the Commission.

In 1959 Congress amended the Communications Act to include the FCC's Fairness Doctrine, which requires (1) that stations devote a reasonable amount of time to controversial issues of public importance and (2) that the stations provide reasonable opportunities for the opposing viewpoints to be heard. The Fairness Doctrine includes a personal attack provision that requires stations to give persons or groups reasonable opportunity to reply when their honesty, character, or integrity has been attacked in a news or public-affairs presentation.

The Fairness Doctrine was upheld by the Supreme Court in 1969 in the famous Red Lion decision concerning radio station WGCB in Red Lion, Pennsylvania. Five years earlier the station had carried a syndicated program with the Rev. Billy James Hargis which contained an attack on a journalist, Fred J. Cook, who asked for reply time and was refused. WGCB maintained that the free speech guarantee of the First Amendment protected the station from claims by those who did not share its opinions.

But the court stated: "When there are substantially more individuals who want to broadcast than there are frequencies to allocate, it is idle to posit an unabridgeable First Amendment right to broadcast comparable to the right of every individual to speak, write or publish." In addition to portraying the public's rights as paramount, the decision also reaffirmed the individual station's responsibility for the programs it carries.

That obligation was established by the FCC in its *1960 Programming Statement* and its *1976 Primer on Ascertainment of Community Problems by Broadcast Applicants*. The latter asserts that long residence in an area is not necessarily evidence of familiarity with community needs. The primer gives guidelines to stations on how to learn about the needs of their communities and how to respond to those needs on the air. The 1976 guidelines have had the effect of making community-need ascertainment an ongoing procedure rather than a ritual performed only at license-renewal time.

The FCC's most basic tool in regulating the broadcast industry is its power to deny license renewals. Broadcast licenses are awarded for three-year periods and are renewable if the FCC continues to deem the licensees qualified. In the fifty-year history of broadcast regulation, relatively few licenses have been denied. In the main, those licenses were lost for illegal acts by the owners, deliberate misrepresentations to the FCC, and serious infractions of the commission's rules and technical requirements. The FCC has never taken away a license on its own initiative for the most important reasons of all: inadequate or inappropriate programming, or a station's failure to serve the public interest, convenience, and necessity.

In practice, license renewal is an almost automatic procedure. Unless an application for renewal is challenged by a citizens group, the application is not likely to be seen by an FCC commissioner. All such license-renewal forms are handled by staff members in the FCC's Broadcast Bureau, whose recommendations are ratified by the commissioners in weekly meetings.

In one important section of the license-renewal form, each broadcaster must describe the amount and nature of the nonentertainment and local programming planned for the forthcoming license period. This "promise" is then measured against the licensee's "performance" at the next renewal time. It is not uncommon for the promise to exceed performance by substantial amounts. But even when a station's record is poor in this regard, all that may happen is that the FCC staff will issue a strongly worded letter requesting an explanation.

The rubber-stamping of license renewals is in part an effect of the commission's personnel limitations; with thousands of applications to be processed annually, the staff would have to be increased severalfold to examine each case thoroughly. Unable to ascertain community needs itself, the FCC has come to rely on feedback from citizens groups, chiefly in the form of petitions to deny the renewal of licenses of stations that have been derelict in their obligations.

Yet the FCC's attitude toward citizen involvement through the years has been far from supportive. Local and national activist groups have had to fight for the commission's adherence to the public-interest standard, to fight on occasion for the enforcement of the commission's own regulations, and to fight for virtually every reform in the system to make broadcasters more responsive to the audience from which they profit handsomely.

In a sense, the FCC has seemed to be intimidated by its own authority, as if denying a license was too extreme a sanction ever to be invoked. Although the commission may, at its discretion, impose fines of any amount, the forfeitures rarely exceed a few hundred dollars—a meaningless penalty for stations that generate many hundreds of thousands of dollars in profits. Moreover, the commission has had a history of being more forceful in its authority with small stations than with the large ones owned by powerful and influential companies.

This reticence to apply harsh economic sanctions derives partly from the FCC's long-standing sense of obligation to the early entrepreneurs who invested huge amounts of capital to build a local and national broadcast service that, for all its faults, has won the appreciation of most of the people.

What the commission tends to overlook is that the public's investment in broadcasting—in its purchase of television and radio sets—exceeds by far the broadcasters' investments in equipment and programming. By 1978, some thirty years after television became a mass medium, the public was estimated to have spent close to $90 billion for television sets and accessories, a figure that does not include the cost of energy, repairs, and installation. The total public investment in television alone is considered to be about twenty times the total expenses of the television industry during the same period.

The FCC is also held in check by its own conflicting mandate. The same Communications Act that empowers the FCC to regulate in the public interest also prohibits the commission from becoming involved with actual program content, lest it practice government censorship. So although programming is the most significant measure of a station's service to the public, the commission generally refrains from passing judgment on programming or from using program-content criteria in making renewals.

Aside from its reputation for timidity and neglect, the FCC is frequently accused of being closer to the industry it regulates than to the public it is supposed to represent. There is something to this. Nearly all station owners are represented by Washington-based communications lawyers, who provide not merely specialized legal counsel but also carefully cultivated pipelines to the FCC commissioners and key members of the staff. The members of the Federal Communications Bar Association make it their business to maintain friendly professional relations with the FCC staffers who, more often than not, make the decisions on broadcast matters which the commissioners will later approve. These relations give the lawyers special opportunities to learn about current FCC policies and thinking and to influence informally the formation of those policies.

The deck is thus stacked against the citizens-action groups, most of which are strangers to FCC procedures and lack, besides, the inside personal contacts. As a result, these groups rarely receive the assistance and attention that broadcasters are accorded through their Washington lawyers.

Moreover, the coziness that exists between the federal agency and the broadcast industry is nurtured by the fact that many who work for the commission—including the commissioners themselves—look to the industries they regulate, or to the Washington law firms, for their next jobs. Of the thirty-three commissioners who served between 1945 and 1970, no fewer than twenty-one later became affiliated one way or another with the communications industry.

Nor can Congress be depended upon to safeguard the public interest where broadcast issues are concerned. Members of Congress are heavily and skillfully lobbied not only by the National Association of Broadcasters in Washington but also by the broadcast interests in their own states. Elected officials are acutely aware of each broadcaster's ability to provide them with valuable media exposure in the years between elections. Friendships are born of this, and few industries can line up congressional support as effectively as broadcasters to defeat legislation that may be threatening to their interest.

The
Audience
and
the
Public

At the television networks, programming decisions are made in New York and creative decisions in Hollywood. This is the reason for considerable executive shuttling between the two coasts, with rarely a stopover in the three thousand miles between. "The public," a network president once astutely observed, "is what we fly over."

Under the business system in which they operate, it seems to make little difference that the networks have hardly any personal contact with the people who make up their audiences. Television's practitioners do keep a close watch on their viewers, but from a distance—the distance of audience ratings and millions of dollars' worth of other statistical research.

From these data, intensively studied by the networks over the years, a portrait of the television audience has emerged. The networks have learned, for example, that although 97 percent of the households in the United States have television, only one third of that potential audience does two thirds of the daily viewing. The medium's viewers divide themselves into three general classifications: steady, habitual watchers of television, for whom being at home inevitably means sitting before the set; occasional watchers, who use television frequently but are not "hooked" on it; and selective watchers, who will tune in for certain broadcasts but otherwise can be indifferent to television.

Playing the sure thing, the networks concentrate on the habitual and occasional viewers and scarcely bother to court the selective group. Thus when the networks speak of the "television audience" they are not referring to the 76 million households equipped with television sets but only of the number that patronizes them regularly or fairly regularly.

Their programming tasks are eased by the knowledge that a predictable number of people will be watching each hour of the night, week after week, regardless of the intrinsic appeal of the shows. During the peak-viewing winter months, some 70 million viewers will unfailingly be before their sets at 7:30 P.M., and their numbers will increase steadily until 9:00 P.M., when they reach 90 to 100 million viewers. After that the viewing levels decline gradually to 50 or 60 million at 11:00 P.M.

These audience levels are achieved automatically, whether there is light entertainment on all channels or reruns or a presidential address. This is why advisers to Presidents try to schedule their important speeches at nine o'clock. (In fact, sets in use generally go up for presidential addresses, since the regular tune-in for television is augmented by selective viewers who switch on their sets especially to hear what the President has to say.)

There are two ways of looking at these statistics. One is to recognize, perhaps with dismay, that about half the population is drawn to television every evening. The other is to take heart in the fact that the other half of the population does not feel compelled to watch without good reason. The half that is watching may be called the audience, but it is both halves that make up the public.

Broadcasters have displayed a tendency to use the terms "audience" and "public" interchangeably, as though they were synonymous. This bit of flawed semantics has fathered the industry's claim that television is a cultural democracy, a medium in which programs succeed or fail by the public vote (the "vote" being expressed, however, in *audience* ratings). It has also created a confusion in the broadcaster's mind about the meaning of "public interest."

A great many broadcasters are satisfied to define the public interest as "what the public is interested in." By that they surely mean "what the audience is interested in." If they really mean the former, then they are admitting their own failure to perform in the public interest and should be asked to explain why they persist in scheduling what consistently does not appeal to half the public.

Admittedly, "the public interest" is a slippery phrase that eludes concrete definition, but what is certain is that, in the broadcast context, it refers to all people and not merely to those who spend the most time with television and radio. Consider how different the Communications Act would be if the key language had been phrased "the audience interest, convenience, and necessity." The late Walter Lippmann, in a wry mood, once offered this definition: "Public interest is what men would do if they thought clearly, decided rationally and acted disinterestedly." The myth that television is a cultural democracy makes cancellations of programs appear to be actions taken in the public interest. But in fact they are almost invariably actions taken in the broadcasters' own economic interest.

Among them, the networks divide billings of close to $3.5 billion a year. Whether ABC, CBS, or NBC gets the largest share of that pot depends on how well each network scores in each half hour, since advertising rates for programs vary according to the popularity of the shows with young adults. Schedules are ripped up and remade with no concern about whether viewers are inconvenienced by the changes, or bereft because of them, but only with the object of fattening the network's ratings averages.

Network programmers do not have to worry about generating audience—as theaters, arenas, and booksellers must—only about capturing the viewers who are sure to be there in their predictable nightly numbers. A network is successful in any time period when it wins 30 percent or more of those tuned to television, since that signifies the attainment of approximately a one-third share of the pie in a three-network competition. Programs receiving less than 30 percent have forfeited audience to the other networks, and those programs become candidates for cancellation.

There is so much money in the television marketplace that no network can be a loser today. In the failure-proof business that television has become, even the shows that get inadequate viewing (inadequate can mean twenty million viewers) will make a profit. The frantic competition among the networks is fueled not by a struggle to survive but by the mania to exceed the rival networks in riches.

In vying for audience, the networks have reduced their program forms to a relative few. The effective program formats for daytime have been found to be game shows and soap operas. Little else may enter this domain. At night the prevailing forms are situation comedies, action melodramas, variety shows, movies, big-time sports, and specials. The depressing mix is almost unalterable. When a network offers a program of greater substance—a documentary or a concert—the regular audience does not switch off the television but instead moves by the millions to the other networks. The penalty for trying to upgrade programming, therefore, is to make one's competitors wealthier, and that is something no network wants to do.

What results from such a system is a homogenization of programming. While in theory three networks and more than seven hundred commercial television stations should make for a great diversity of program services, in practice they add up to only a single type of service, in triplicate. The networks (and their stations) compete for the same audience under the same terms, play by the same rules, buy programs from the same sources, and interchange their personnel. Indeed, if all programs in a prime-time schedule were juggled and redistributed among the networks at random, ABC, CBS, and NBC would not be radically different from what they are today.

Public television does offer a viewing alternative, but it is too weak a force to break the networks' lockstep. It represents negligible competition for viewers. The most popular weekly series on public television—whether *Upstairs, Downstairs; Great Performances;* or *Monty*

Python's Flying Circus—attracts an audience far smaller than the lowest-rated shows on ABC, CBS, and NBC. In part this is because public television does not reach the entire country; many of its stations transmit weak signals, and two thirds of them are on the UHF band, making them somewhat harder to receive than commercial stations. But there is also another reason: Public television draws much of its audience from the ranks of selective viewers.

Researchers at the commercial networks make a distinction between people who watch *television* and people who watch *programs*. It is a most important distinction. People who watch television ask themselves, "I wonder what's on today," and then settle in for an afternoon or evening, accepting what is offered and allowing themselves to be carried from one show to the next. These dedicated viewers of television are not deterred by the removal of their favorite shows. People who watch programs behave quite differently. They tune in purposefully for specific broadcasts and switch off the set when the program is over or, if they are dissatisfied with it, even earlier. They have come not to watch anything but to watch *something*. Putting it in other terms, some surrender their leisure time to television and become addicted to the tube as a mental anesthetic; others use the medium discriminately, making conscious and independent choices, often frivolous ones.

Children can be taught at an early age that coming home from school or play and switching on the television for whatever might be there is not the same as deciding in advance what they wish to watch—even if it must be a rerun of *Gilligan's Island*—and turning on the set at the appropriate hours. The selections children make will improve with time, but it is critically important that they learn first to make selections.

The public that network executives fly over is the docile, easily manipulated mass audience that watches TV because that is the easiest thing to do in waking life. The rest of the public is not enchanted by flickering images alone, has other ways to spend its time, and is so differentiated it cannot be known through mere statistical research.

The Meaning of "Public" in Public Broadcasting

What commercial broadcasting has failed to be, public broadcasting was supposed to become—a system devoted to excellence and to the exploration of the full range of programming possibilities, without bondage to the marketplace. As in such countries as England and Japan, where a beneficial friction exists between commercial and public systems to the betterment of both, an American public broadcasting institution was envisioned as a strong second force on the airways that would not only provide viewing alternatives but would also, by its challenge, spur commercial broadcasting to higher achievement.

Public television and radio, when they were created by the Public Broadcasting Act of 1967, were to be noncommercial entities in the service of the people. The word public was not used with reference to government, as it is in the terms "public sector" or "public funds," but rather was meant to be synonymous with "people." Although the financing for the system was to come partially from congressional appropriations, there were to be no political strings attached.

To insulate the system from political influence, and to distribute federal funds and provide parental leadership for the new industry, the Corporation for Public Broadcasting (CPB) was created as a private, nonprofit organization. In turn, the CPB created the Public Broadcasting Service (PBS) to manage the traffic of national programs to the newly interconnected stations.

Undeniably, with its menu of cultural programming and topflight British dramas, public television satisfies many viewers, who express their appreciation by happily contributing financial support to local stations in their fund-raising drives. But overall, public television has fallen far short in realizing the ideals that were enunciated for it at its birth. It has been accused by minorities of elitism and discrimination, it has frustrated independent producers who have found access difficult, and it has flirted several times with performing in the service of the government—or more specifically, the Administration in power. Its dependency on corporate underwriting for national programming verges on commercialism, and its reliance on British television for its most important fare verges on scandal. Finally, its on-the-air appeals for funds have become even more obnoxious forms of huckstering than the basest commercial spots on the other channels.

Moreover, far from becoming a strong second force in broadcasting, the public television industry has appeared content to remain a weak imitator of commercial TV. Thus, instead of creating the healthy friction that would stimulate an increase of public-service programming on the commercial channels, the existence of public television has served to curtail public-service activity and to relieve commercial broadcasters of the pressures to provide such programming. In many markets around the country, commercial stations have yielded the responsibility for educational or other limited-interest programming to the local public television channel.

The seriousness of this consequence is compounded by the fact that few public stations devote much time to local programming. According to a 1976 survey of the industry by the CPB, only around 10 percent of the average public station's programming was locally produced—a shabby record for an industry sworn to the doctrine of "grass-roots localism." Even commercial stations that are loaded to the hilt with network service and syndicated reruns do better than that.

Another CPB study, conducted over a period of eighteen months and concluded in the fall of 1978, found public broadcasting seriously delinquent in serving the needs of minority people—Blacks, Hispanics, Asians, and Native Americans—both in programming and in employment practices. The study found that only 9.4 percent of the total $5.8 million worth of programming financed by the CPB in fiscal 1977 were programs made by, or of specific interest to, minorities. A sampling of local public television stations showed that each invested less than five thousand dollars a year to buy national programming for minorities. Meanwhile, National Public Radio had allocated only 3 percent of its $2.7 million program budget that year to serve specialized audiences.

The same study showed that only 6 public radio stations out of some 184 in the continental United States and only one public television station (WETV, Atlanta) out of the 272 on the air were controlled by minorities.

As for the industry's employment record, the report showed that minorities constituted 14 percent of the 10,865 full-time public broadcast employees. Fifty-one percent of public radio licensees and 16 percent of public television licensees had no minority employees at all. In the top three job categories—officials, managers, and professionals—59 percent of the radio licensees and 33 percent of the television licensees had no minority staff members. Although public television had a higher percentage of minority employees than did public radio, the percentage of minorities in management positions was higher in radio than in television.

The study concluded that the short shrift given minority-interest programming by public broadcasting was the result of "inadequate minority participation in program decision-making at the national public broadcast level."

Actually the reasons go well beyond that, and they bear also on the elitism of the noncommercial system. Many of the stations derive a large portion of their discretionary funds from the financial contributions of viewers. It is customary for them to court the audience that would most likely be contributors—affluent middle-class people who enjoy the arts and are interested also in gourmet cooking, gardening, science, and national issues. To program for the downtrodden minorities, therefore, is to lose an opportunity to raise money.

On the national level, where programs need corporate or foundation underwriting to get on the schedule, it is often the private corporations (Exxon, Mobil, Atlantic Richfield, and others that would be considered advertisers in commercial television) that determine what PBS will carry. If the business and public-image purposes of these companies are not served by particular types of programs—controversial documentaries and minority-interest broadcasts, for example—then those programs stand little chance of getting on the air. So much for public television as the people's television.

What went wrong between the dream for public broadcasting and the reality? Officials of the industry tend to lay the blame for the unfulfilled promises entirely on the lack of adequate funding for public broadcasting in this country. The Public Broadcasting Act did indeed hobble the system it created by its failure to make provisions for the permanent financing of noncommercial television and radio, either by a dedicated tax or by some other means. But money is only one of several reasons public broadcasting has not become what E.B. White, for one, hoped it would be: "our Lyceum, our Chautauqua, our Minsky's and our Camelot."

The problem probably began with the allocation of stations. Public broadcasting was an afterthought in America; it was allowed to make its way only after commercial radio and television were firmly entrenched. This was the reverse of the pattern in most other advanced nations, where commercial television was not authorized until the public broadcasting system had developed fully.

When the FCC's four-year freeze on television licenses ended in 1952, the commission—on prodding by Commissioner Frieda Hennock, the Ford Foundation, and a number of activist educators—reserved 242 channels for noncommercial television dedicated to education. Later the number of allocations was increased, but the fatal flaw in the plan was that two thirds of the stations for noncommercial use were on the UHF band. Thus the system was condemned to be weak before it began.

UHF, the band encompassing channels 14 to 83 on the television set, still cannot be received in many households, and in many others the signals come in poorly. Even people who can receive UHF well are not inclined to seek out programs on it because the tuning involves more effort than the tuning of VHF stations. As a result, even with seventy more stations on the air than any of the commercial networks, PBS is unavailable to a large portion of the population. In such key cities as Los Angeles and Washington, public broadcasting struggles for an audience because its outlets are on UHF.

There has been some gradual improvement of the situation. With the help of the all-channel receiver law of 1962 that requires all sets to be manufactured with UHF capability, and with cable TV and other technological developments, public television has slowly been overcoming its UHF handicap. But other problems of the system are harder to eradicate. Most of them can be traced to the almost forcible thrusting of the public broadcasting precepts upon an existing system, known up to the passage of the Public Broadcasting Act as educational broadcasting.

The noncommercial stations that went on the air between 1953 and 1967 had an educational mission. Some were licensed to colleges and universities, some to state authorities or boards of education, some to municipal educational agencies, and the remainder—mostly in the larger cities—to nonprofit civic corporations. The stations were not interconnected by land lines until 1967, and each grew up as an independent fiefdom governed by a board of directors that was typically white, male, representative of the local power establishment, and essentially conservative in its approach to programming.

These stations were willing enough in 1967 to form a loose confederation for purposes of securing government or other financing, and they were even willing to adopt the new label of *public* television. But they would not easily alter their system of values or their primary educational mission. Nor were they willing to accept the hegemony of a national organization or to have their programming dictated by the Ford Foundation or community-owned big-city stations. Today the boards of many public stations remain unrepresentative of the entire community, especially where low-income minorities are concerned, and they still cling to elite notions of not having to make overtures to a general public.

Largely because the stations are more concerned with their own sovereignty than with what would benefit a national system of public broadcasting, funds are scarce for ambitious national programming, standards for program acceptance are uneven around the country, and public participation at a national level is all but meaningless.

Finally, the flaw in the system is that it is not insulated by the CPB from government interference. The fifteen members of the corporation are nominated by the President and must be approved by the Senate. In addition, President Richard Nixon introduced the practice of making the corporation's board a dumping ground for patronage appointees. This is not only a dubious way of achieving political insulation but also a poor, if not bizarre, way of providing the system with parental governance.

If it had not been notorious for Watergate, the Nixon Administration would have been notorious for its sometimes successful efforts to control public broadcasting and reshape it into a system that did the bidding of the executive branch. Documents from the files of the White House Office of Telecommunications Policy for the years 1969 to 1974 show how the Nixon Administration purged public television of commentators it felt were hostile to the President, then went to work to rid it permanently of news, commentary, and public-affairs programming. The records are mostly memorandums between Clay T. Whitehead, director of the Office, and high-ranking White House staff members, including John D. Ehrlichman, domestic affairs adviser; Charles W. Colson, special counsel to the President; Peter M. Flanigan, assistant to the President; and H.R. Haldeman, the chief of staff. They and the

President were obsessed with the belief that public broadcasting had a liberal political tilt. They frequently referred to "anti-Administration content" that Mr. Nixon thought dominated noncommercial programming. The documents were obtained by *The New York Times* through a Freedom of Information Act inquiry.

Whitehead began the attack at a convention of public broadcasters in Miami in the fall of 1971. He admonished the station operators not to pursue aspirations of becoming a "fourth network," expressed the Administration's displeasure with public television's liberal leanings, and called for the system to deemphasize national programming and embrace what he called "grass-roots localism." Moreover, he said, a broadcast system that received financial support from the government had no business dealing in news and public affairs.

His remarks were given force some months later when President Nixon vetoed a $165-million two-year funding bill for public broadcasting, making it plain that he considered the industry undeserving of the money because its power had become too centralized.

On another front, Flanigan launched an effort to subvert the Corporation for Public Broadcasting. In a memorandum to the President dated June 18, 1971 he argued that the Administration could not kill CPB, because of its support from educators and its "generally favorable image." He therefore recommended "restructuring" the corporation to stop its support of liberal causes, restraining it in such a way that in the future it "will be difficult for other administrations to alter."

Nixon was not able to control a majority of the CPB before he was forced to resign in 1974. But he controlled four board members who met regularly with White House officials to keep them apprised of CPB policy trends and programs that were being considered. One of the four was Thomas B. Curtis, a former Republican Representative from Missouri, who became the Administration's handpicked chairman of the board. He then operated as if he were a member of the White House staff. He was seconded by Henry Loomis, the president (i.e., paid executive) of the corporation, who met with Whitehead before every board meeting and leaked confidential information to him about what programs CPB might finance.

Whitehead, who is now president of the Hughes Satellite Corporation, mounted a campaign to smear the putatively liberal newsmen Sander Vanocur and Robert Mac-Neil. He explained his tactics in a November 24, 1971 memorandum to Haldeman:

> After Vanocur and MacNeil were announced in late September, we planted with the trade press the idea that their obvious liberal bias would reflect adversely on public television. We encouraged other trade journals and the general press to focus attention on the Vanocur appointment. Public television stations throughout the country were unhappy that once again they were being given programs from Washington and New York without participating in the decisions. My speech criticizing the increasing centralization of public television received wide coverage and has widened the credibility gap between the local stations and C.P.B. and NPACT [National Public Affairs Center for Television].

He went on to tell of the next plan to "quietly solicit critical articles regarding Vanocur's salary coming from public funds." A corollary to the plan was to "quietly encourage station managers throughout the country to put pressure on NPACT and C.P.B. to put balance in their programming or risk the possibility of local stations not carrying these programs. Our credibility on funding with the local stations is essential to this effort."

Hungry for the federal dollar, public broadcasting bowed to the Nixon pressure. It promptly decentralized and proceeded to fulfill amply a Whitehead prophecy embodied in a memorandum to Mr. Nixon on November 15, 1971: "We stand to gain substantially from an increase in the power of the local stations. They are generally less liberal and more concerned with education than with controversial national affairs. Further, a decentralized system would have far less influence and be far less attractive to social activists." Exploiting the divisive issues that plague relationships in public broadcasting, he said, "provides an opportunity to further our philosophical and political objectives for public broadcasting without appearing to be politically motivated."

The power of the Public Broadcasting Service as a network was dissolved and replaced by a station-by-station voting mechanism called the Station Program Cooperative. Localism became the byword of public television. Vanocur and MacNeil soon lost their contracts, and the noted conservative William F. Buckley Jr. wound up with a regular series. News, commentary, and public affairs were generally played down.*

Just before his forced resignation, Nixon grudgingly signed the five-year authorization bill that Whitehead had promised the public broadcasting industry if it complied with the Administration's wishes. Nixon still saw "serious dangers in the existence of a Federally funded broadcasting network," according to Whitehead.

Bill Moyers later observed that this kind of political interference was in some ways inevitable. He said:

> Uncle Sam isn't a benevolent philanthropist. When he gives you an allowance he expects to be invited to Sunday dinner at least. Anyone who says you can get money from Congress or the White House without strings attached must have been reading a 1950s high school civics book. No way . . . regardless of the party.

Can the public ever expect more favorable treatment from public broadcasting? Perhaps. The Public Telecommunications Act of 1978, enacted by Congress in November, offers some hope. It is meant to make public broadcasting more public than it has been. Under the act, public stations are required to open their board meetings to observation by the general public. Each station is required to create a public advisory board that represents a cross section of the community the station is licensed to serve. Stations were given 180 days from the bill's passage—or until May 2, 1979—to assemble their public ad-

---

* MacNeil has since returned in prime time with the Monday-to-Friday *MacNeil-Lehrer Report,* but it does not present his views. *Report* is an interview program on issues and people in the news, with opposing opinions strictly balanced.

visory boards and to begin conducting open board meetings. The only meetings that may be closed to the public are those involving contract negotiations and matters of utmost confidentiality, such as personnel changes.

The new act also authorizes three years of federal financing for noncommercial television and radio in amounts totaling $600 million. Under the law, all stations receiving federal funds distributed by the Corporation for Public Broadcasting must conduct their policymaking meetings in the open, place their annual audits in a public file, provide access to independent producers, and encourage public participation through community advisory boards. In addition, they must be in compliance with the Equal Employment Opportunity requirements.

Public broadcasting entities that do not meet the conditions of the act will not qualify for government grants. The requirements for openness apply also to the national programming services, the Public Broadcasting Service, and National Public Radio.

Officials of the industry, conceding that public broadcasters have a history of using federal funds for maintaining their physical plants, interpret the new law as an attempt by Congress to channel more of the money into programming through more direct involvement of the public. The congressional subcommittees that drafted the legislation made it clear that the bill was also intended to correct public broadcasting's tilt toward the more affluent people in the audience.

With the built-in mechanisms for greater openness, public accountability, public access, and financial responsibility, the legislation authorizes $180 million for public broadcasting for fiscal 1981, $200 million for 1982, and $220 million for 1983.

In January 1979 the Carnegie Commission on the Future of Public Broadcasting concluded an eighteen-month study of the industry with a report that said, "We find public broadcasting's financial, organizational and creative structure fundamentally flawed. There is little likelihood that public television and radio might consistently achieve programming excellence under the present circumstances." The commission recommended that the system be reorganized, beginning with the elimination of the Corporation for Public Broadcasting and its replacement by a better-insulated body to be called the Public Telecommunications Trust. This proposed organization would watch over the system, evaluate performance, and serve as fiduciary agent, but it would have no voice whatever in program decisions. Those decisions, on a national level, would be made by another new organization, the Program Services Endowment, whose fifteen-member board would consist of accomplished artists, educators, producers, journalists, and broadcasters. The Endowment, which would control millions of dollars, would not be a network; it would work only at generating programs with no authority to impose its programs on the stations in the system. If stations were to reject certain programs, those works would be offered to other new media, such as cable television or video disks.

To make this new and improved public broadcasting system work, the Carnegie Commission recommended annual funding of $1.2 billion a year, with about half to come from the federal government and the remainder from the present variety of nonfederal sources. To reduce the government's burden, the commission proposed that commercial broadcasters and other occupants of the electromagnetic spectrum be charged annual fees for the use of the public airwaves. These fees were expected to raise about $200 million annually.

The Carnegie Commission's report was a call for a national commitment to public broadcasting, a commitment that would recognize it as a "major cultural institution that can play a decisive role in bringing together the pluralistic voices and interests of the American community."

But even if the Carnegie Commission's recommendations should be turned into enacted legislation, the noncommercial station operators are likely to continue to go their separate ways in their differing philosophies, leaving public broadcasting to remain a name without a concept. Whether public broadcasting is to grow to become an important second force on the airwaves or is merely to continue as a secondary broadcasting service will depend on more than increasing infusions of money. It will depend on a shared sensibility by autonomous stations. The system is still young enough and still pliant enough to be shaped by the needs and desires of the American people.

Comes
the
Revolution

Dazzling new kinds of television are entering the picture, the gifts of an exploding technology. Individually or in combination, the electronic marvels that are now bidding for a place in the scheme of national and global communications could touch off a second television revolution, one that might deal a severe jolt to the existing commercial television system and profoundly affect the ways consumers use the medium.

Through these new delivery systems, with their special components and antennas, the ordinary television set may take the place of the bygone neighborhood movie house and could become, as well, an extension of the opera house, football stadium, library, university, classroom, church, town council, and hospital emergency room. And we can foresee the set serving as a burglar- and fire-alarm system, a home computer, and a receiving unit for electronic mail.

The promises of the new media are intoxicating, but so also was the promise of conventional television in the 1940s. It is not too early to begin the watch on the emerging media, while public policy for them is still being

formed. Here are the principal new technological developments:

Pay cable is pay television by means of cable TV, which
results in the consumer paying first for basic cable service
and then an additional monthly fee for the premium fare,
usually a channel of movies, sports events, and special en-
tertainments. The growing consumer demand for pay
cable is highly significant, since it has been spurring the
expansion of cable TV to the densely populated areas of
the country. Modern cable systems typically offer thirty
active channels of television, although some provide up to
thirty-six. It follows that more channels open the way to
more varied forms of programming and greater public access. By the fall of 1978, approximately 18 percent of U.S.
households subscribed to cable TV. Media experts predict
that when the penetration reaches 30 percent, cable TV
will become a national advertising medium and a true
competitor to conventional television.

Communications satellites are the space-age alternative to telephone land lines as a means of interconnecting
television stations or cable-TV systems for the distribution
of programs nationwide. Two separate commercial systems—Western Union's Westar and RCA's Satcom, each
with two satellites aloft—operate as common carriers,
making their transponders available to all comers on a
first-come-first-served basis. The Public Broadcasting Service already distributes all its national programming by
satellite, as do the various burgeoning cable networks and
such special program services as the Christian Broadcasting Network and the Hughes Sports Network. The key to
the expansion of satellite transmission is the proliferation
of the special receiving antennas known as earth stations.
Their increase is assured by the rapidly dropping prices for
the compact 4.5-meter models to serve local TV stations
and cable systems. The combination of new cable channels and easy, economical national distribution by satellite will inevitably add up to new networks.

Satellite-to-home broadcasting is a development that
permits households to receive programming directly from
a satellite, without its being relayed by a local station,
through the use of a small parabolic antenna that might

be mounted on a rooftop or hung outside a window. Direct-to-home satellite transmission is in limited use in Japan and has been demonstrated successfully in this country, but the FCC is not likely to authorize it for the United States because of the political pressures that would be brought to bear. Broadcasters fear the competition and have already effectively argued that local stations might be killed off and that the doctrine of localism would therefore be jeopardized.

Fiber optics is the technology that converts electronic signals to lightwaves, transmitting them over thin glass fibers with the use of a laser beam. Optical fibers are so thin that a small bundle of them can carry more than one thousand channels of television. They are likely in time to take the place of telephone wires and cable TV's coaxial cable; in fact, a single "wire" of optical fibers can provide both telephone and cable-TV service.

Electronic data transmission systems permit viewers to call up on the screen a wide range of printed material from an information bank through the use of special decoders. Some forms operate through cable and some through conventional television in conjunction with the telephone. Initially their application is in the retrieval of news, stock reports, and other current information, but ultimately they may be tied in with computer data banks.

Home video recorders are mass-marketed video-tape consoles that permit consumers to record up to four hours of programs off the air to be watched at the viewer's discretion. These units can also play prerecorded tapes purchased in a store.

Video disk players are the video counterpart of the phonograph, whose program material comes on twelve-inch metallized records. One type uses a stylus and a grooved disk, another a laser beam and a smooth disk. The disks contain thirty minutes of programming on each side.

Two-way cable is cable TV with a return wire that permits interactive communication. The viewer at home, equipped with a unit containing response buttons, may send digital signals back to the transmission center. When a computer is joined to the system, the interactive mode becomes useful for public-opinion polls, purchases of specific pay-television fare, viewer participation in game shows, the ordering of merchandise advertised on television, and burglar- and fire-alarm protection for the home.

Late in 1977, in the test market of Columbus, Ohio, Warner Communications began operation of a remarkable two-way cable system to which it gave the trade name Qube (pronounced *cube*). The significance of the name is that the system is multifaceted, involving the use by viewers of a hand-held console with four sets of buttons.

Ten buttons in one column serve as the controls for over-the-air television. These buttons can bring forth on the screen any of ten conventional television stations, those of the immediate locale and others imported from distant cities. A second row of ten buttons is for pay programs— movies ranging in price from $1.50 to $3.50 each, sports events, college courses, and cultural events—each priced on a per-viewing basis.

The third row of ten buttons provides special channels programmed by Warner, each channel dedicated to a single type of fare—for example, high-grade children's shows (many of which are imported from abroad), news, sports, religion, and culture. The principal channel in this row offers continuous live local programming, all of it designed to elicit viewer responses.

These responses are implemented by the last row of five buttons, which are larger than the others. By pressing the appropriate buttons, people at home may vote on performances in an amateur variety show and answer questions in quiz programs. They may also purchase products from the screen advertised by local retailers or bid on merchandise being auctioned. Of greatest consequence, however, is the use of the response buttons to poll viewers on local and national issues, with the results tabulated by the computer in less than a minute. On occasion, during an interview with a public figure, viewers have been asked whether they believed what the subject was saying. The response from the audience can put the interviewee on the spot in a way that does not happen in conventional television.

Many people have a tendency to look to this new panoply of electronic miracles for the solutions to the problems presented by conventional television. Indeed, a number of government policymakers have been doing just that. They reason that a great proliferation of channels as promised by cable and fiber optics relieves the problem of spectrum scarcity. New networks via the satellite figure to break the competitive lockstep of ABC, CBS, and NBC and end their creative stagnation. Video recorders, the video disk, and pay television liberate the consumer from the agendas set by the networks and local stations and let the viewer become the programmer. Interactive television permits instant feedback and direct viewer participation.

All this is true—technology does come to the rescue—but there are caveats. The timetable for the next television revolution (if, indeed, there should be one) is highly uncertain. The present system should not be deregulated, as some in Washington propose, purely on an assumption that technology will change everything tomorrow.

The new media have yet to prove themselves in the marketplace, and there is every chance that some will fall into the technological backwash, along with the Picturephone, 3-D movies, and facsimile—all of which held great promise in the 1950s but failed to become thriving businesses.

Moreover, these media come freighted with an array of new public problems that may themselves beg for regulatory solutions. Qube, for example, gives viewers the power to answer back, but it is still Warner Communications, a giant media and entertainment conglomerate, that controls the questions. And all the new television delivery systems will be in the hands of business concerns anxious to improve their profit status every year. So there are no guarantees that the promised cultural and educational fare, and the programs of social significance, will last any longer in these new media than they did in commercial broadcasting.

In Washington the sentiment is strong to open the doors wide to these new technological developments, to remove all regulatory obstructions so that they can have every opportunity to compete with each other and with conventional television in the marketplace. This desire, in part, grows from the knowledge that for almost thirty years the FCC, responding to the commercial broadcast lobby, had stifled the development of over-the-air pay television and slowed the growth of cable TV with burdensome regulations for nearly as long.

But there is another reason for the wish to turn the new technology loose. A number of key figures in government have become disillusioned with the broadcasting system spawned by the Communications Act of 1934, and they see no point in trying to preserve it in its present form. Recognizing that a sham has been made of the principle of licensing and that the ideals of the act have never been realized, these officials are prepared to write the whole thing off as a failure and to let market forces decide what shape American television is to take.

Representative Lionel Van Deerlin, Democrat of California and chairman of the House Communications Subcommittee, introduced in 1978 a radical bill to revise the Communications Act from top to bottom. The bill envisioned a wide-open marketplace for all the various forms of electronic communications—including conventional television and radio—so that all might compete evenly for survival, in the spirit of free enterprise. This meant taking virtually all the regulatory wraps off radio and, after a ten-year period, television as well. Conspicuously absent from the proposed new version of the Communications Act was the phrase "public interest, convenience and necessity." The public-interest standard that had been the foundation of American broadcasting for half a century was to be junked.

The bill was not a piece of mischief by friends of the commercial broadcasting establishment. It was born of realistic appraisals and honest concerns about the existing system, chiefly by political liberals. The position of the bill's proponents, gained from interviews with them, may be summarized in this manner:

> Let's do away with the pretense that we can have a system of broadcasting in the public interest, and let's be coldly realistic. Broadcasters are business-people whose natural drive is to maximize profits. Isn't it pointless, and wasteful of energies, to ask them to act against their true nature? Isn't it an ex-

ercise in futility to try to convince them that they are public trustees who have to serve the public interest ahead of their own economic interests? Why not let market forces come into play, as they do with the print media, ultimately providing something for every taste and need? Why not let the marketplace (i.e., the consumers) and not the government decide how these media should perform and which ones should prevail? Competition is in the public interest, and this is the surest way to achieve diversity and public service.

Congressman Van Deerlin's proposal found adherents among influential persons in government who were staunch upholders of the First Amendment. These officials had always been uneasy about broadcast regulation. In their view the awarding of broadcast licenses by a government agency, the withholding of licenses, and the stipulation of conditions under which such licenses are granted are all abridgments of the freedom of expression guaranteed by the First Amendment. The print media are free from all such regulation. When different standards of freedom are applied to different media performing similar services, they concluded, the First Amendment itself is in jeopardy. The idea of a new Communications Act that would do away with license renewals appealed to them because they could see no reason why the free-speech amendment should be so dangerously compromised for a public-interest standard that is largely myth.

But the Van Deerlin bill's detractors far outnumbered its supporters. The freeing of commercial broadcasting from the public-interest standard alarmed leaders of the broadcast reform movement and galvanized all the various citizens groups into a single opposing force. Everett C. Parker, director of the Office of Communication, United Church of Christ, condemned the bill as "the greatest giveaway of the public's rights since Teapot Dome." In his testimony at the subcommittee hearings on the bill, Dr. Parker stated:

> In stripping away all requirements for licensees to account for how they serve the public, the bill grants

television, radio and other telecommunications monopolies the right to determine, unilaterally, what political, economic, cultural and educational ideas shall be disseminated to the American public. Such a privilege has never before been conceded in our society to a monopoly. It is the exact power that the Declaration of Independence denied to the King of England.

Others pointed out that the bill ignored all the progress that had been made within the system during the previous decade, the period since the public began to play an active part in broadcast affairs. The fact that the present system had failed, they argued, did not necessarily mean it was a failure. Semantically, a failure is a concluded flop, but to say something has failed is still to allow for the chance that it can be made to work. The achievements of the citizens movement, they contended, were evidence that the system can be made to work and that the ideals of the 1934 Communications Act could still be realized.

Among opponents of the bill were commercial television broadcasters, which made this one of the rare times that they and the consumer groups were on the same side of an issue. Most broadcasters objected to the bill because it proposed to charge them fees for the use of the electromagnetic spectrum in exchange for relieving them of their public-interest responsibilities; it was the fee that bothered them, not the principle. But a considerable number of broadcasters were also seriously troubled by the prospect of being freed from the public-interest standard, since it was clear to them how that would affect their industry.

Some predicted privately that deregulated television stations would drift fairly quickly into pornography. It was a certainty, they said, that the station most desperate for an audience in its community would turn to explicit sex. If that station succeeded, the rival stations would eventually have to follow. Something like that has been happening to motion picture theaters.

Television in Italy provides the example to illustrate that theory. After years of having only a state-run network on the air, the Italian government in 1976 permitted local commercial stations to go on the air as unregulated entities. To gain attention, one began showing hard-core adult movies like *Deep Throat*. Another instituted a game show on which a housewife removed articles of her clothing when contestants gave wrong answers and put them back on, one at a time, when the contestants' answers were correct. Inspired by the popularity of that program, another station eliminated the game-show gimmick and just brought on women who performed amateur stripteases. Very soon the commercial stations made nudity and pornography an everyday part of Italian television.

Aside from the likelihood of that happening here in an unregulated system, American television stations with no accountability to the public would be free to run violent cartoon programs directed at children, programs offensive to minorities, and editorials that are extreme or even bigoted in their views. They might even elect to provide no service whatever to persons over the age of fifty, since the older adults carry a low priority with advertisers.

The Van Deerlin bill never made it to the floor of Congress in its original form but instead was withdrawn for extensive revision, thanks partly to the lobbying efforts of citizens groups with individual members of the subcommittee. Nevertheless, the bill in retrospect had considerable value. For one thing, it illuminated the need for an enlightened, updated national policy to embrace all forms of electronic communications. For another, it charged up the citizens-action movement, making those groups and others more appreciative of the public-interest standard and more conscious of the need to preserve it. Finally, it brought forward some new assumptions for thoughtful examination.

Early in 1979 Congressman Van Deerlin introduced a new version of his bill with "sweeteners" to make it more palatable to broadcasters and common carriers such as the Telephone Company, and even less attractive to the public-at-large. In the Senate, Senator Ernest F. Hollings, who chairs the Senate Communications Subcommittee, has submitted a bill aimed mostly at reform of common carrier regulation, but which would provide for substantially less regulatory oversight for radio and for a five-year license term for television.

One central assumption of the Van Deerlin proposal is

that two-way cable TV, the video disk, and the other new media will be as pervasive in this century as commercial broadcasting and that they will all be able to compete in the marketplace on an equal footing. This assumption overlooks the fact that the existing television system has had a long head start on the new developments, is firmly entrenched with a large and generally satisfied audience, and therefore will have a decided advantage in an open market.

No matter how many new channels and networks come into being through cable TV and satellites, the existing commercial stations will continue to dominate viewing for a long time for two principal reasons: First, they are the established Main Streets of television and have conditioned viewers to believe they are where the action is; second, they involve what most people regard as a "free" service, while the newer channels will be charging fees. (It should be noted that people do pay for so-called free television through increased prices for the advertised products.)

The competition, therefore, will be between new media drawing many thousands of viewers, and conventional television drawing tens of millions of viewers.

Another assumption of the Van Deerlin bill is that the vigorous competition among the various electronic media will make for a vast, groaning board of fare, just as competition in the print media has produced a wide range of books, magazines, and periodicals. What is more likely, however, is that most of the entrepreneurs will take aim at the most lucrative markets to exploit, and for the most part these markets will be penetrated by entertainment rather than by information. An indication of that is pay cable's concentration on movies and sports, types of fare already offered abundantly by commercial television.

Advocates of deregulation believe that by granting full First Amendment protection to broadcasters and cable operators the cause of robust, wide-open debate on major issues will be served. They anticipate a renaissance of hard-hitting investigative journalism. But there is more reason to believe that those who control the television signals, relieved of their public trustee obligations, will concentrate on programs that have the greatest profit potential and will shun low-rated public affairs programming. Investigative journalism is expensive—more expensive for the size of the audience it delivers than a situation comedy.

As attractive as the new media may be, they were not developed on huge amounts of venture capital for eleemosynary purposes. Already it is evident that there is a lively market in the video-cassette field and in the Qube system in Columbus for what euphemistically are called erotic movies, while no such lively consumer market has as yet developed for instructional programming. Like other corporations, those that are marketing the new electronic developments have a mandate to increase their profits every year, and this inevitably means that their initial good intentions with educational and cultural programming will give way steadily to more lucrative software.

Moreover, if today the local television or radio broadcaster is considered to be a powerful media figure in a community, the operators of cable systems may become thirty times more powerful, since they will control the flow of signals on thirty channels and not just one.

Because there is the potential for such extraordinary control of mass communications in cable TV, many public-interest advocates have been campaigning for a national policy that would designate cable systems as common carriers. A common carrier, exemplified by telephone and telegraph, is a medium that is open to use by the entire public on a nondiscriminatory basis and at posted rates. This would mean that cable operators would control the wires and the transmitting mechanism but would have no involvement whatsoever with the programming. Under such an arrangement, anyone who wished to broadcast would have the opportunity to buy a portion of television time for his or her own programs and would be able to sell advertising in those shows just as television stations do.

The cable industry opposes common-carrier status if for no other reason than that it would involve federal supervision of the rate structures and a ceiling on profits. Cable entrepreneurs point out that Warner Communications would never have risked $20 million on a futuristic system such as Qube—indeed would not have bothered to invent the system—if it were so divorced from programming that it could not determine what would be offered.

It is impossible not to be impressed with Qube and with its potential to provide programming for every taste and need in a community. Beyond that, the great cultural change that is likely to occur when people are able to use television interactively, instead of merely watching the screen passively, is altogether desirable. Yet, at the same time, it is difficult not to be concerned about the possible social consequences of this new form of television. A computer checking out television households every six seconds and making records of how people vote in polls, or of what programs they choose to watch, clearly invades privacy. This is the Big Brotherism that had been prophesied by George Orwell in his famous novel, *1984*.

Charles D. Ferris, the chairman of the FCC, once remarked that he feared the implications for politics of Qube's ability to conduct instantaneous public-opinion polls, for although it is good for legislative representatives of a community to know where their constituents stand on an issue, there is also a danger that the polls may be conducted before the public is fully informed on the issues. Elected officials might endanger their careers if their votes on legislation did not follow the mandate of the Qube poll, even though their own clear understanding of the issues might dictate a different vote.

While the viewer may take comfort in having the ability to interact with television by sending forward his or her responses on issues, the fact remains that the cable company controls the questions and may shape them according to its own political or ideological purposes. Further, in a cable-computer opinion poll the viewer is left to take it on faith that the result flashed on the screen will be the result shown by the computer. Like a sighted person playing heads-or-tails with a blind person, the cable company is free to deceive by reporting a false tally on issues in which it has an interest and thus may influence legislation as it chooses.

But of all the reservations about the new media, the

most serious one concerns ownership. Satellites, pay-cable networks, and sophisticated interactive systems such as Qube are big businesses, far too big for independent entrepreneurs or small groups of local investors. Knowledgeable observers predict that if Qube proves hugely successful in Columbus, and if Warner begins to expand it around the country, other entertainment and media conglomerates will quickly enter the field by gobbling up existing cable systems across the country and converting them to systems resembling Qube. As the game gets richer, it will require the capital investment that can be made by companies like Gulf & Western, Time Inc., MCA Inc., and General Electric. Thus, the powerful in the media stand to become more powerful still.

Along with this prospective development is the possibility of American Telephone & Telegraph Inc. extending its telephone monopoly to cable TV. This could happen because the technologies of television, telephone, and computers have all converged into a single technology, and because AT&T has already begun the conversion of its installations from copper wire to fiber optics. Since the same bundle of optical fibers that provides telephone service may also provide cable-TV service, it becomes unnecessary to have two separate wires entering the home.

Under present regulations telephone companies are prohibited from owning cable-TV systems. But there is talk in Washington of lifting the ban, because telephone companies could solve the problem of bringing adequate television service to remote rural areas of the country where cable companies cannot afford to build systems. In fact, the Van Deerlin bill proposes to give telephone companies the freedom to enter the cable field.

So in a time when a concerned public is calling for more diversified ownership that is more broadly representative of all segments of the society, the looming trend is for more concentrated ownership. It is a mistake, then, to trust the emerging electronic media, as Representative Van Deerlin does, to bring about the reforms in conventional broadcasting that citizens groups have been working toward. If anything, these media are raising new issues and posing new threats to the public interest.

The
Real
Critics
of
Television

Newspaper critics, writing about television every day, can take no credit for the important changes that have occurred in broadcasting in recent years or for the heightened responsiveness of broadcasters to the people they are obliged to serve. Over the years, the press on the whole has treated television as a wellspring of personality news and gossip and has confined its journalistic scrutiny of the medium to evaluations of the new programs that come on the air.

The real critics of television and radio—and the only effective ones—are the citizens-action groups. These organizations began to spring up in the mid-1960s, spontaneously, out of the growing concern by private citizens for the industry's apparent indifference to the effects these enormously powerful media were having upon their children, their neighbors, and society as a whole. It was also these citizens groups—and not the press—that finally spread the decades-old news that because the airwaves belong to the people the public has rights in radio and television. This news unfortunately has spread slowly. Most of the population has still not been advised.

Ordinary people understandably feel powerless today

before giant corporations and great bureaucracies. Yet when armed with the powerful force of morality, some knowledge of what the law provides, and the courage to speak out, a single person can make the earth shake under the mighty. The action of one private citizen, John F. Banzhaf III, a New York lawyer, set in motion the developments that led to the banning of more than $200 million worth of cigarette advertising on television and radio in 1971. It all began in 1966, when Banzhaf requested free time from New York station WCBS-TV to answer its cigarette commercials under the Fairness Doctrine. When the station rejected the request, Banzhaf filed a complaint with the FCC. His case was based on the fact that the Surgeon General's report proved conclusively that cigarettes were a health hazard and that therefore the promotion of smoking on TV and radio was a controversial issue of public importance. The FCC upheld the complaint and ruled that all stations would have to broadcast one antismoking spot for every three cigarette commercials. A few years later, Congress decided that media licensed in the public interest had no business promoting products that were clearly not in the public interest, and it passed the Public Health Cigarette Smoking Act of 1969, banning cigarette advertising on broadcast stations. The ban did not extend to newspapers, magazines, and billboards because they are not licensed in the public interest.

Around the same time, four women raising families in the Boston suburb of Newtonville became so alarmed at the excess of violence and hard-sell commercials in the networks' Saturday morning children's programming that they pooled kitchen funds to form an organization called Action for Children's Television (ACT). In 1970 these women presented the FCC with a petition that carefully documented what ACT believed to be exploitation and abuse of children by broadcasters and advertisers. ACT supported its charges with meticulously researched program logs and with the expert testimony of pediatricians and child psychologists. The petition called for a number of specific reforms in children's programming, including reductions in the amount of advertising carried in these shows, the prohibition of certain advertising practices that were taking advantage of children's gullibility, and the elimination of violence.

Broadcasters at first paid little attention to ACT and so were stunned when the FCC acknowledged the seriousness of the issue and gave notice of a proposed ruling, inviting comments from the industry and the public. The ACT petition drew an unprecedented one hundred thousand letters of support, mostly from parents, which helped offset the tremendous industry pressure for the FCC to keep its hands off this lucrative portion of the broadcasters' schedule.

ACT has grown into a national organization with chapters in most major cities. As a result of its continuing efforts, the networks have striven to improve the quality of their children's programming. They have upgraded the role of the children's programming executive in the company and have sought the consultation of educators and child-development specialists. Meanwhile, both the FCC and the Federal Trade Commission have been kept busy considering new issues raised by ACT and by other organizations that keep watch on television's treatment of the young.

But the crucial instance of citizens action, the landmark case that gave not only impetus but also weaponry to the broadcast reform movement, was the one which caused the Lamar Life Insurance Company to lose its license to operate WLBT-TV, Jackson, Mississippi. This case, which began in 1964 and was consummated in the courts in 1969, determined among other things that citizens have "standing" to participate in commission rule-making and renewal hearings. It also established that ordinary people —citizens of a community—could file petitions to deny the renewal of a broadcast license.

During its first thirty years the FCC had been willing to listen to the complaints of public groups about broadcast station performance, but it insisted that only the commissioners themselves were authorized to represent the public in FCC proceedings. Only "parties in interest" had the "standing" with the commission needed to file petitions in opposition to a license renewal. A "party in interest" was defined by the FCC as someone who could prove that he or she suffered "economic injury" or "electrical interference" from a station.

In 1964 the Office of Communication of the United Church of Christ was joined by two black citizens in Jackson, Mississippi—Robert L.T. Smith and Aaron Henry—in a petition to deny a license renewal for Lamar. The petition, painstakingly researched and documented, charged WLBT with blatant racial and religious discrimination and with excessive commercialism. Charges were backed up with a written record of station practices derived from monitoring, which detailed WLBT's deliberate exclusion of blacks from its programming and its refusal to air anything but the segregationist view during the civil rights movement of the 1960s. A station licensed to a city whose population was 42 percent black was openly refusing to serve the needs and interests of the black community.

While acknowledging the seriousness of the charge, the FCC said it was unable to recognize the petitioners as "parties in interest" and concluded that justice would be served by extending WLBT's license for a probationary period of one year instead of granting the customary three-year renewal.

The petitioners challenged the commission's refusal to recognize them in the U.S. Court of Appeals. In 1966 a three-judge panel unanimously ruled that citizens groups did indeed have "standing" as consumers with a genuine and legitimate interest in a community station's broadcast service. The FCC was ordered to hold a renewal hearing for WLBT in which citizens groups could participate. In the decision written by Chief Justice Warren Burger (then a circuit judge) the court said:

> The Commission has always viewed its regulatory duties as guided if not limited by our national tradition that public response is the most reliable test of ideas and performance in broadcasting as in most areas of life. The Commission view is that we have traditionally depended on this public reaction rather than on some form of governmental supervision or "censorship" mechanism.

Justice Burger's opinion went on to say that "experience demonstrates consumers are generally among the best vindicators of the public interest. In order to safeguard the public interest in broadcasting, therefore, we hold that some 'audience participation' must be allowed in license renewal proceedings." The decision noted further that broadcasting is not "a purely private enterprise like a newspaper or an automobile agency" and that when a broadcast franchise is accepted it may not be operated at the whim and caprice of its owners the way a newspaper can, but instead "is burdened by enforceable public obligations." Three months after issuing the decision, Justice Burger amended it with this acerbic statement: "After nearly five decades of operation the broadcast industry does not seem to have grasped the simple fact that a broadcast license is a public trust subject to termination for breach of duty."

Following the court's instruction, the FCC conducted a two-year investigation of WLBT and then held a hearing. When all that was over, the FCC in 1968 granted Lamar a full three-year renewal of the license. The petitioners once more appealed to the court, and again the court unanimously reversed the commission. This decision, which also was written by Justice Burger (and announced on the morning of the day he was sworn in as Chief Justice of the United States) ordered Lamar to "vacate" WLBT and directed the commission to invite new applicants for the license.

The station was allowed to remain on the air and to be operated by a racially integrated nonprofit caretaker group until the FCC decided upon a new licensee, a decision that has not yet been made fifteen years after the petition to deny renewal of the license was filed.

The WLBT case opened the way for the local citizenry to express itself in broadcast matters. It established the right of citizens to participate in FCC proceedings and at the same time proved the effectiveness of the "petition to deny" as an instrument of change. A petition to deny differs from a license challenge. The latter is made by persons who want to become broadcasters themselves by winning away a station license from an incumbent. The petition to deny has no acquisitive motive; it is an assertion by members of the community that the broadcaster has violated its public trust or is providing inadequate service and therefore is unworthy of the privilege of operating a television or radio station there.

The WLBT decision came the year before the Supreme Court's Red Lion decision, which tackled the prickly First Amendment issue and concluded: "It is the right of the viewers and listeners, not the right of broadcasters, which is paramount." The Red Lion landmark decision also said:

> It does not violate the First Amendment to treat licensees given the privilege of using scarce radio frequencies as proxies for the entire community, obligated to give suitable time and attention to matters of great public concern. To condition the granting or renewal of licenses on a willingness to present representative community views on controversial issues is consistent with the ends and purposes of those constitutional provisions of the press. Congress need not stand idly by and permit those with licenses to ignore the problems which beset the people or to exclude from the airways anything but their own views of fundamental questions.

With the WLBT and Red Lion decisions the desirability of broadcaster-community interaction was certified. Citizens now had unquestionable rights they were being encouraged to use, and also some powerful tools with which to effect change within a lethargic regulatory system.

But filing a petition to deny is still a drastic measure for citizens groups to take, given the FCC's reluctance to take away broadcast licenses. During 1976 nearly 180 such petitions were filed, and the FCC—whose makeup at the time was decidedly conservative—rejected all of them. Before embarking on a petitioning effort, a public-interest group would do well to try a simpler means of gaining results. Direct negotiations with stations—some would call it "jawboning"—has in a large number of instances proved productive.

Citizens' action in broadcasting becomes most effective when a number of diverse groups coalesce around an issue. Alliances of organizations for ethnic minorities, feminists, and civil rights advocates, for example, have had marked success in altering employment practices at stations. As a result of such united efforts, broadcasting is no longer the all-male, lily-white industry it was before 1970, although there is still a long way for it to go to become fully representative of the society. The inclusion of women, blacks, Hispanic-Americans, Native Americans, Asian-Americans, and other minorities in the executive ranks is an assurance that programming will be more diverse and more expressive of the concerns of the entire public.

While it is possible for a single person or a single group with a properly documented complaint to be heard at a station or at the FCC, there is obviously more impact and a better chance for success when the numbers are larger and more broadly representative of the community. Station managers usually do whatever they can to discourage petitions against their licenses, even though the odds for renewal are in their favor. A petition to deny subjects a station to a costly legal defense and adds substantially to the work load of the management and staff. Therefore, when important local groups are part of a coalition that asks to be heard at a station, the station manager cannot afford to ignore them.

One of the classic cases of successful negotiations occurred in Texarkana, Texas with television station KTAL. In 1968 the Office of Communication of the United Church of Christ aided twelve black groups in Texarkana in filing a petition to deny the license renewal of KTAL. As with WLBT, the station was charged with egregious racial discrimination both in its broadcasts and in its hiring practices. Besides, KTAL had moved its main studios more than fifty miles away, so as to favor the more profitable market of Shreveport, Louisiana, which meant a curtailment of its coverage of Texarkana, the community it was licensed to serve.

Before the FCC could decide to schedule a hearing, the station agreed to meet with the petitioners to negotiate. The parties arrived at a settlement that substantially improved KTAL's employment of blacks and its service to the Texarkana area. The FCC approved the agreement as binding, and it encouraged other citizens groups to consider negotiated settlements as alternatives to petitions to deny for the redressing of grievances. Since 1968 more than one hundred stations have concluded similar agreements with the FCC's blessings.

A negotiated settlement will be disqualified, however, if it should abridge a broadcaster's responsibility for what is broadcast. The FCC knocked down a 1972 settlement between Los Angeles station KTTV and the locally based National Association for Better Broadcasting (NABB) because the agreement had the effect of the citizens group dictating what should be put on the air. In negotiations following a petition to deny, the station had agreed not to broadcast a number of children's series which the NABB had listed as violent. The commission refused to accept the agreement, maintaining that it constituted a form of program censorship and that KTTV was delegating some of its programming authority to an outside group.

Sometimes in their zeal to achieve positive ends, citizens groups resort to questionable means. As a result, issues that might be resolved by the healthy interaction of the broadcaster and its community are instead resolved nefariously by censorship. The use of threats and pressure is one of the pitfalls of citizen involvement, and it is one that can be damaging to the movement as a whole.

In the late 1970s several groups, including the National Parent-Teacher Association, claimed credit for driving violence off the networks by threatening to boycott the advertisers in programs that featured violence. Certain other groups took aim at the advertisers whose spots appeared in programs that were deemed to be excessively permissive with sex. While there is certainly nothing reprehensible in viewers venting their displeasure with the corporations that sell their products through television—indeed, stockholder resolutions are to be encouraged—the technique of using boycott campaigns against advertisers as a means of killing off shows has dangerous implications.

Stockholder resolutions, such as those that have been spearheaded by Stewart Hoover of the Church of the Brethren, involve efforts by shareholders in large corporations to dissuade those companies from supporting television programs that may be harmful to the society and to adopt affirmative policies for television advertising. This activity is quite different from boycott campaigns, whose dangers are not theoretical but have already been experienced.

In the McCarthyist climate of the early 1950s, an organization known as Aware Inc. successfully engineered a blacklist of persons whom the organization considered to be sympathetic to the Communist enemy, usually on the basis of scant information. Advertisers—who spend fortunes to create customers for their products, not to make enemies—were warned that they would be boycotted if they sponsored programs that employed blacklisted people. The networks cravenly went to work to weed out those "undesirables" to make life more comfortable for the advertisers. Dozens of programs fell to the Aware campaign, and hundreds of careers were blighted, most of them unjustly.

More recently, the advertiser-boycott technique was used by special-interest groups in attempts to prevent the broadcasting of two network specials. In one instance, pressures exerted by the country's gun lobby caused virtually all advertisers to quit a CBS news documentary, "The Guns of Autumn," which gun enthusiasts considered unfavorable to their interests. Despite the lack of advertising support, the program went on as scheduled. But the message of the campaign was not lost on the networks, and it is likely to chill all future thoughts of airing programs that may, by implication, favor gun control.

In the second instance, a number of fundamentalist religious groups tried to block NBC's telecast of "Jesus of Nazareth" by threatening to boycott the prospective sponsor, General Motors. General Motors had purchased the American rights to the 6-hour-and-37-minute film, which had been directed by Franco Zeffirelli and co-produced by Britain's ATV and Italy's RAI-TV. Led by Bob Jones III, president of Bob Jones University, the religious groups protested the showing without ever having seen the film. They based their objection to it on a magazine interview with Zeffirelli, in which he indicated that Jesus would be portrayed as an ordinary human in the film biography. The program was televised nevertheless, but by that time General Motors had been scared off, and Procter & Gamble became the sponsor. The program drew a huge rating and will probably become an annual network offering for the Easter season. If the boycott campaign had been a success, the film might never have been aired in the United States.

The campaigns to boycott advertisers whose spots appeared most frequently in violent programming were, in a sense, unfair. Except for a few specials, most programs are not "sponsored" by advertisers as they used to be. Instead, programs carry spot announcements that have no identification with the shows themselves. Advertising time on a network is typically purchased by computer today; an advertiser indicates how much it wants to spend and the kinds of people it wishes to reach, and the network's computer responds by printing out a schedule indicating where the spots would run.

This practice somewhat resembles the way advertisers buy space in magazines and newspapers. In the print media the advertisement is totally divorced from the editorial content of the articles surrounding it. The same is intended in television. Yet if the consumer boycott can work in television under those circumstances, what is to prevent the technique from being extended to print by groups that may wish to suppress certain articles or political points of view?

There is no question that the tactic of harassing advertisers can produce results in the media, but the achievements are of dubious value if their by-product is censorship.

The system provides other recourse. Groups concerned about the excesses of violence or sexual titillation on a network should attempt to deal with the problem at the local affiliated station. The local broadcaster, who is legally responsible for everything it puts on the air whether or not the program is sent out by a network, may then be made to explain why it is *its* judgment that these excesses serve the best interests of the community. The campaign might follow this blueprint:

1. *Document the case.* A few weeks spent in monitoring the station should produce some hard evidence to make

the complaint credible. The evidence might show that 70 or 80 percent of the peak-viewing hours in a typical week were devoted to programs of a violent or sexually explicit nature. If these were indeed the statistics, it is possible that the broadcaster itself hadn't realized how imbalanced the schedule was. The monitoring study might be backed by other research. A demographic breakdown of the audience for the station might find that 20 percent of the viewers in the prime-time hours were elderly persons. This fact could be pertinent to the complaint in light of behavioral studies which show that violence on television affects the elderly by making them afraid to leave their homes. It goes without saying that a significant number of the viewers of violent and sexually exploitative programs are children.

2. *Enlist the support of other groups.* The issue is one that might be of concern both to people on the political right and to people on the political left. Church groups, local PTAs, local political officials, legislators' spouses, teachers, students, business people, unions, women's groups, ethnic organizations, environmentalists, and any or all of the broadcast-oriented citizens-action groups might wish to become part of the coalition on the issue. In a pluralistic society, numbers are eloquent and broad representation persuasive. They make it difficult for the broadcaster to dismiss the complaint as a pleading by an insignificant minority of elitists who essentially hate television.

3. *Advise the press of what is going on,* so that the larger public might know. Although TV critics on the whole do not keep watch on the industry as they should, they know a good story when they see one, especially when it is backed with documentation and research. An army of important local groups complaining to a station about an overload of exploitative programming is likely also to draw the attention of editorial writers.

Placed in the position of having to answer to its community for the excesses, the broadcaster is likely to cite ratings that reveal the objectionable shows to be popular with much of the audience. The obvious response is that popularity is not a measure of the public interest, any more than the popularity of drugs with young people is proof that drug use is in their best interest.

If, say, twenty stations around the country were faced with determined protests from citizens, they would quickly convey the message to their networks that they could no longer accept programs that will jeopardize their licenses. The networks cannot afford to put on programs that a significant number of stations refuse to carry, because that would impair their advertising prospects. Soon enough the objectionable programs would vanish, not because of censorship but because the marketplace could not support them.

One
Innocent
Step
Behind

At the midpoint of the 1978-79 television season, the three networks wheeled in twenty-two new weekly series as replacements for their failures—a record number. But bigger than the rising rate of program fatalities was the fact that not one of the twenty-two new entries was designed to trade on violence or sexual titillation. What made this new policy a matter of extraordinary interest was that it represented a turning away from flavorings that had pervaded prime-time television for two decades. For the first time since the invasion of the Westerns in the late 1950s, networks that were striving to improve their positions in the ratings did not resort to sex or violence as subjects.

Why did all three networks decide at the same time to travel a new road? What prompted the swing to a more wholesome kind of fare? Some network officials, such as Gene F. Jankowski, president of CBS Broadcast Group, attributed it merely to the cyclical nature of popular television programming. He suggested that it happened because sex and violence had run their course as inducements to viewing. That explanation might have sufficed if it were not for the fact that such sex-oriented series as *Three's Company, Soap,* and *Charlie's Angels,* and such hard-action shows as *Starsky and Hutch* and *Hawaii Five-O,* were still ranking very high in the ratings.

Clearly, there was still an *audience* to exploit for sex and violence when the networks changed their direction. What brought on the change, apparently, was the *public outcry* against the exploitation. It was a cry that built up over twenty years, as consumerists, religious groups, politicians, social scientists, psychiatrists and eventually the American Medical Association and the National Parent-Teachers Association (NPTA) all began to register concern. As the numbers grew, the protest became impossible for the networks to ignore.

Moreover, in what would have to be considered desperation, some of the groups, including the NPTA, resorted to pressure tactics upon television advertisers, holding them accountable for the programs in which their ads appeared. Sears, Roebuck was one of the notable targets of an organized campaign, one that carried the threat of a boycott. It caused the company to adopt new advertising policies, specifically excluding programs that might be considered offensive for their preoccupation with sex. Other advertisers were driven out of violence-oriented programs by a campaign of the National Citizens Committee for Broadcasting, which held them up to public disgrace. Thus, as certain types of programs were increasingly eschewed by advertisers, their value to the networks declined.

"Organized criticism of TV has itself become a growth industry," Robert Mulholland, president of NBC-TV, complained in a public forum.

> Never before have we experienced the kind of sophisticated, professional public relations and political tactics used by these groups. . . . Determined organizations seem to be on the march—all in different directions. And in most cases these groups conduct their campaigns in ignorance of—or without concern for—the economics of TV, its operating realities, or the structure of a mass medium based on advertising, program suppliers, news responsibilities, affiliated stations and networks.

Fred Silverman, Mulholland's boss as president of the National Broadcasting Company and also the shaper of the company's policies, was less accusative. He conceded in a *TV Guide* interview that NBC would not deal in sex-oriented programs because of public disapproval, and he said in an earlier speech to network affiliates:

> There is a basis for criticism of the television medium. It raises questions as to whether television is doing all it can to realize its enormous potential. . . . We must avoid material that would alienate significant segments of the audience, and we must continue steadfastly to make the difficult and delicate judgments that draw the line between the offensive and the acceptable, under changing public attitudes.

The twin issues of sex and violence—they are justly linked, because they are interchangeable avenues to the quest for an advantage—have been a prime source of public involvement in television from the time the eye-poking Three Stooges raised the concern of parents, and the plunging necklines of Faye Emerson and the busty sweaters of Dagmar upset the puritanical. The involvement has grown over the years from letters of protest to the FCC and members of Congress, to organized campaigns to do something about it, whether picketing a network or pressuring advertisers.

But this is where concerned people come close to trampling on the First Amendment rights of broadcasters, producers, performers, and writers. It is one thing for people to react vocally to ideas and productions they deplore, but it is quite another matter when interest groups, even when they believe they represent the majority, try to bar program material from being broadcast. One is an act of free expression, the other an attempt at censorship.

Although they may seem to bear a resemblance, there is a difference between citizens action in broadcasting, and interference with programming by special-interest groups. Sometimes it is as hard to draw the fine line between them as it can be to distinguish between morality and immorality.

The argument of this book is that it is desirable for people to take an active part in the television and radio processes, since the public-interest standard implies not only that citizens have the right to do so but also that they were intended from the first to be part of the broadcast equation. But it is not desirable—and has never been in the United States—to take any action whose result would be to silence voices or vanquish the journalistic or creative spirit.

In a letter to *The New York Times* in 1978, Eric Bentley defended the rights of people who with political motives attempted to influence the judgments of theater critics against a play about Paul Robeson. He also defended the right of New York's Cardinal Spellman in the 1960s to advise Catholics not to see the play *The Deputy*. Bentley wrote (and the emphases are his):

> It is not only permissible under our laws to try to influence people with words that one hopes are persuasive, it is a *good thing* to try to do this. One *should* try to do this.
>
> The people who considered the Robeson play pernicious not only had the right to say so—in paid ads, in private letters, in all peaceful and normal channels—they had an *obligation* to say so, an obligation to Paul Robeson as they saw him, an obligation to our society as they understand it.

The operative phrase here is "peaceful and normal channels." Bentley was not advocating the suppression of viewpoints or artistic works (specifically the Robeson play) but rather was encouraging just the opposite: free and robust debate on public issues and questions of art. People who feel the need to protect have a duty to make themselves heard.

Elected officials were not indifferent to the public protests against gratuitous violence on television, but Congress lacked the legal means to curb excesses of murder and mayhem on the home screen, being powerless to act in any censorial way. A regulatory ban on television violence was possible only if there were conclusive proof that such programming had broadly harmful social effects. With such proof, the case could have been made that violent television programs were not in the public interest and that therefore, like cigarette advertising, it was appropriate to drive them off the air.

Over the years behavioral studies on the effects of television violence were conducted by the score but had a way of canceling each other out. Although many studies found excessive viewing of violence to be generally harmful, other studies found it was not. In the late 1960s, spurred by the unrelenting war on violence and sex on TV by Rhode Island Senator John O. Pastore, the Surgeon General's office began a three-year study to determine whether a causal relationship existed between the watching of violence on television and subsequent aggressive behavior.

The five-volume report on the $8 million study couched its conclusions in such cautious, ambiguous language that the net effect was inconclusive. Newspaper interpretations of the report gave the impression that it found no relationship between television violence and aggressive social behavior. However, in subsequent hearings before the Senate Communications Subcommittee, conducted by Senator Pastore, each member of the surgeon general's twelve-person advisory committee acknowledged finding that the viewing of violence caused some people to behave more violently.*

Senator Pastore's 1972 hearings were only the latest in a long line of formal inquiries by congressional bodies and commissions into the question of television's effects on social behavior. The first was held in 1954 by Senator Estes Kefauver, well before the invasion of the Westerns on television. Then came the hearings in 1961 and 1964 by Senator Thomas Dodd. Four years later, hearings by the National Commission on the Causes and Prevention of Violence, led by Milton Eisenhower, brother of the former President, sought to determine the role of the mass media in the increase of crime. In 1969 came Senator Pastore's

---

* The Surgeon General's report has been criticized because the television industry was invited to participate in selecting the advisory committee. To insure industry collaboration in the research to be sponsored, the National Institute of Mental Health sent a list of forty candidates for the committee to the three networks and the National Association of Broadcasters. The networks vetoed seven of the forty candidates, so all seven were dropped from the list. CBS, however, did not exercise the veto option. See Kurt Lang, "Pitfalls and Politics in Commissioned Policy Research," in Irving Louis Horowitz, ed., *The Use and Abuse of Social Science* (New York: E.P. Dutton & Co., 1971). Also see Philip M. Boffey and John Walsh, "Study of TV Violence: Seven Top Researchers Blackballed from Panel," *Science*, May 22, 1970.

questioning of Surgeon General William H. Stewart that led to the three-year study.

The television networks responded to each of these inquiries by immediately cutting back on violent programming. But like the driver who slows down when he sights a police car in his rearview mirror, and then accelerates when the law enforcers leave the highway, the networks' judiciousness with violence was always short-lived.

Violence came into the medium on a large scale in the late 1950s when the Hollywood movie studios began producing weekly series, chiefly Westerns, for the networks. This was just about the time—after the quiz-show scandals—that advertisers were ceasing to become sole sponsors of weekly programs, leaving the networks more free to seek larger audiences with any program form and any programming device which served that purpose.

The Hollywood horse operas lent themselves readily to the audience-attracting motifs of action and violence. For example, one show opened with a "tease"—that is, a preview scene—in which the screaming head of a bound man was threatened with being pushed into a bonfire by the heel of his menacer. As the networks began scheduling Westerns to compete with other Westerns, it became a matter of survival for the programs to try to outdo one another in depictions of violence. *The Untouchables*, a police-gangster series set in the prohibition era, came to the medium in 1959 by way of ABC—a network then desperate for ratings points—and unabashedly exploited the viewer appetite for brutality and murder. Its success made violence irresistible to other network program executives and their suppliers in Hollywood.

Meanwhile, something similar was happening on Saturday mornings, where the networks were just beginning to compete with one another through blocks of children's programming. The networks found that as they increased the use of animated cartoons oriented to violence—with monsters, humanoids, and assorted other grotesques—they were rewarded with larger audience ratings and greater profits.

In prime time, one action genre gave way to another.

The Westerns yielded to the private-eye cycle. When that ended, the spy-adventure cycle began. After the Dodd hearings, which produced evidence that network executives had been ordering producers to heighten the violence for the sake of ratings, the networks attempted in earnest to scale down the use of gunplay and bloodshed. But then came a new programming development—the networks began purchasing recently made theatrical movies for prime time—and the good intentions went by the boards.

The influence of the movies on television programming after 1965 cannot be overstated. When movies fairly fresh from theatrical release first appeared on the networks, they devastated all programs in competition. On the whole they were better produced, better written, and cast with bigger stars than any television series. Made originally for an industry that was not licensed to serve the public interest, the movies had far greater artistic freedom than any program produced for television. By 1966 each of the networks had two movie showcases a week. Television producers, finding it impossible to compete with motion pictures, pleaded with the networks to relax their standards for program acceptance so that TV shows might meet the challenge of the movies on a more equal footing. That permission was granted. Thus, whatever was considered acceptable for airing in movies then had to be acceptable for programs produced for television. No double standard. Violence made a powerful comeback.

But the impact of movies on television programming did not end there. Motion pictures became further intertwined with television strategies when, around 1968, the broadcast industry began to find itself in need of artistic and thematic guideposts if it hoped to reach the viewers who would maximize its profits. Up to that time the television business revolved on ratings that measured how many people were watching each show. Advertisers, however, determined that some people were more worth reaching than others, and this began their preoccupation with demographics—breakdowns of the audience according to age and sex. It was the youth market that was prized, because young adults had discretionary money to spend or were setting up new households. They, more than young children or older persons, would be faced with making choices in floor waxes, toothpastes, and cake mixes.

Thus, television advertising rates began to arrange themselves on a scale that was highest for viewers in the age group of eighteen to thirty-four, next highest for those ages eighteen to forty-nine, and lowest for programs that drew predominantly children or the elderly. The problem for television was that its most devoted audience was made up of people who were under eighteen or over fifty. The elusive group in between—the demographic group the advertisers craved—were the people who more or less regularly went to the movies.

To win those people, the motion picture industry had to offer what television could not offer: pictures that dealt explicitly with nudity and sex, that used street language without inhibition, and that took violence to new extremes. To win the same people, television eased its standards and showed many of those movies, editing out the nudity, bleeping out some of the words, and blunting some of the violence. But each time the boundaries of acceptance were stretched for a movie, they remained stretched for the rest of television's fare.

Thus began a bizarre cycle—the motion picture industry staying ahead of television to keep its audience, and television imitating what succeeded in movies, with the rationale that the moral values of the country were changing and that it was merely keeping up with the times. Indeed, network presidents repeatedly pointed out that they were following and not leading, and that they were always an innocent step behind the other media.

Movies continue to set the pace and to provide the inspiration for new television programs. But that would not be the case if the movies were patronized mainly by juveniles and senior citizens. If young adults ages eighteen to forty-nine were spending their time at antiques fairs and dog shows, television—which professes an obligation to reflect reality—would be an innocent step behind and dogs and antiques would be in prime time.

As for the interesting development in the middle of the 1978-79 season, when all three networks shunned sex and violence in their replacement shows, no experienced observer of the industry expected this to be a permanent reform. History had made it easy to predict that when the irate public was mollified, the network furthest behind in the ratings would mount a surprise attack on its rivals with new programs laced with violence and sexual frankness, capture an audience, and start the cycle all over again.

Beyond
the
Ratings

When patience runs out, the choice today is between shooting the television set or joining the broadcaster-community dialogue as a voice to be heard beyond the ratings. People who are content to accept television and radio passively, foregoing their right to participate in the regulatory process, are condemned to remain merely the merchandise of those media—manipulated, packaged for the market, and sold on a cost-per-thousand basis like herds of cattle.

By the irony of how the system has evolved, it has become the public—and not the broadcaster or the FCC—that determines by its vigilance whether radio and television operate in the public interest. Because that is the case, citizens groups have become a full-time component of the American broadcasting system. They occupy the void between a reluctant FCC and an essentially amoral broadcast industry distracted by the ever-blooming opportunities to increase profits.

The citizens movement, sometimes referred to as the broadcast-reform movement, by now is so established that it even has a structure. At the heart of it are the hundreds of disparate local groups—some seeking rights for minorities, others a fair shake for women, children, homosexuals, or the elderly; most of the remainder are organized as watchdog groups on ideological, moral, journalistic, and consumerist issues.

Many of these groups maintain contact with one another and exchange information, and when their interests converge on an issue they join forces as a coalition. In addition to these full-time organizations, there are ad hoc groups that come into being when radio stations threaten to change formats from classical music to rock or when the sensitivity of some local issues calls for a media watch.

Some local groups concentrate on public radio and television—either as "friends" of the stations to assist in fund raising, or as critics of station policies they believe give short shrift to the less affluent members of the community. There are also groups that are concerned with local standards being adopted for future cable-TV systems or that serve as gadflies to existing cable systems.

Counterpart organizations exist on the national level. Among the more prominent ones are ACT (Action for Children's Television), the National Black Media Coalition, AIM (Accuracy In Media), the National Association for Better Broadcasting, the National Citizens Committee for Broadcasting, and the media subsidiaries of the Gray Panthers, the National Organization of Women, and the National Gay Task Force. The Cable Television Information Center, based in Washington, D.C., provides guidance and information to local groups that keep watch over cable TV.

Leaders of the broadcast-reform movement include such national religious organizations as the Office of Communication of the United Church of Christ, UNDA-USA (Catholic broadcasters), the Unitarian-Universalist Association, the United Methodist Church, the National Council of the Churches of Christ in the U.S.A., and the National Conference of Catholic Bishops.

Another vital facet of the movement are the public-interest law firms that specialize in broadcast cases and represent or advise, on a *pro bono* basis, many of the individuals and groups that are trying to redress grievances with local broadcasters. These firms include the Citizens Communications Center, the UCLA Communications Law Program, the Media Access Project, and the Center for Law and Social Policy.

Most activities of the national citizens groups and public-interest law firms are financed by private philanthropic foundations. Among the most active in this area are the Ford, Markle, and Rockefeller family foundations, the Veatch Program of the North Shore Unitarian Society, Plandome, New York, and the Carnegie Corporation of New York.

Completing the structure are large national organizations that do not exist primarily to deal with broadcast issues but are intensely interested in some of them. Among those that have played forceful roles in broadcast reform from time to time have been the National Parent-Teacher Association, the National Council of Churches, the American Civil Liberties Union, Consumers Union, the League of Women Voters, the National Association for the Advancement of Colored People, and the Urban League.

If the concerns of all these varied groups had to be expressed in a single word, the word would be *access*. The control of radio and television by a relative handful of business executives—many of them monopolizing the local media through ownership of newspapers and cable-TV systems, as well as TV and radio stations—has left large segments of the society frustrated by their inability to use the airwaves like those who have been awarded the privilege. It is not vanity that motivates them to seek access but rather a sense of outrage over the injustices that sometimes are perpetrated by broadcasters.

Black people, for example, rarely appeared on television before 1968 except in sporting events, as tap dancers on the *Ed Sullivan Show*, or as waiters, porters, maids, and savages in other weekly series. The one black series that did exist in the early days of television was *Amos 'n' Andy*, a program that offended blacks because it patronized the race and reinforced demeaning stereotypes. Eventually it was driven off the air by black pressure groups.

This egregious exclusion of blacks from television, in

the face of a public-interest standard that requires broadcasters to consider all significant groups in making their programming decisions, was justified by the industry on economic grounds. The networks explained that their affiliates in the South would "pull the plug"—that is, not carry the network transmission—if they were to send out a program in which blacks were prominently featured. Moreover, the networks said, if stations refused to carry the show, then advertisers would refuse to sponsor it.

Indeed, they were able to cite a shameful example in the *Nat (King) Cole Show*. In 1957, NBC attempted to build a weekly variety series around the popular recording artist, and many white entertainers offered to make guest appearances free, to help the show succeed. But advertisers did decline sponsorship for fear of boycotts in the South—this was before the civil rights movement—and the show was canceled in its first season.

Tyrone Brown, a black FCC commissioner, made a strong and moving case for the retention of the public-interest standard in his testimony to the House Communications Subcommittee on the proposed rewrite of the Communications Act in 1978. Commissioner Brown said:

> I vividly recall my early experience when virtually all people I saw on television were white. With Commission prodding things have improved, but I for one would not want to rely solely on the altruism of broadcasters to air programming reflecting the world as it really is. My sons watch TV. I want them to grow up watching a more realistic representation of our diverse society than I did as a child.

The long-held industry assumption that many white families would not welcome black people into their homes through television went up in smoke when the networks—spurred by the WLBT decision—began to offer such series as *Julia, Sanford and Son, The Jeffersons, Good Times,* and *What's Happening.* Each of these series ranked for several years among the ten most popular shows in the country. Then came *Roots,* the twelve-hour limited series tracing the history of an American black family from the enslavement of its progenitors. The TV adaptation of Alex Haley's best-selling book broke all viewing records when it was presented by ABC on eight consecutive nights in

1977. *Roots, Sanford and Son*, and the others might never have come to television, and the medium might have remained lily-white to this day, if citizens had merely sat mute and obediently accepted what was being served up by the networks, stations, and advertisers.

Women, homosexuals, the elderly, Italian-American organizations, and other ethnic groups have also organized to counter the television stereotypes that were damaging their lives while contributing to network and station profits. The access sought by these groups was not necessarily access to the airwaves but access to the minds and consciences of those who produce and select programming.

Local groups of Japanese-Americans and German-Americans—for the sake of their children—had to plead with television stations not to show the low-budgeted patriotic movies that were made during World War II, which portrayed the Japanese and Germans as grotesquely evil and made free use of pejorative nicknames. Mexican-Americans were forced to make organized appeals to stations and advertisers to cease being made the butt of jokes. It was through the efforts of these groups that the comic Frito Bandito commercial was removed from the air. The advertisement was only one of numerous instances in which the Mexican accent was presented as amusing and the Mexican male as either a gaudy bandit, not to be trusted, or as a sombrero-wearing idler on permanent siesta.

Meanwhile, consumerists and environmentalists were nettled at being unable to reply to commercials that violated their standards and beliefs. Serious writers, satirists, and independent documentary producers began a push for access to a television medium that had become a closed club of network programmers, Hollywood studios, and advertising agencies. The drive for artistic access extended also to public television.

Fringe political candidates and ideological groups—on both the left and the right of the spectrum—resented being shut out of radio and television, and they organized to find ways to make their views heard. The conservative and influential Accuracy In Media is but the largest of a number of watchdog groups on broadcast journalism looking for what they consider to be deliberate distortions or inaccuracies in the news.

In addition to voicing their views to the broadcasters and applying pressure on them, the various citizens groups create a general awareness of the issues that concern them by holding conferences and publishing newsletters and magazines. Many also serve to counterbalance the powerful broadcast lobby by testifying at FCC and congressional hearings. It was primarily the citizens lobby that defeated attempts in Congress to repeal the Fairness Doctrine and to extend the three-year broadcast license to a period of five years. These were well-organized efforts involving large coalitions of national and local groups, careful legal analyses of the proposals, and extensive discussion with Representatives, Senators, and members of the subcommittee staffs. In addition, large numbers of interested citizens turned out to testify at hearings before the Senate Communications Subcommittee, building up the case that the bills would adversely affect the public interest.

As might be expected, citizens groups are resented by the industry as nuisances, troublemakers, hysterics, and threats to the business stability of radio and television. Most broadcasters understandably prefer the old way, when people sat quietly by, not realizing they had rights in the electronic media, or else expressed their rage by destroying their television sets in the privacy of their homes.

Yet there are more than a few thoughtful broadcasters in the country who concede that citizens groups have been, overall, a positive influence on the industry, helping to improve broadcasting if in no other way than by identifying the worst malefactors among station operators. Some suggest also that it is probably healthy for broadcast business executives to feel the constant presence of their most severe critics in the community.

The citizens movement has realized scores of victories —from improving the quality of children's programming and advertising to upgrading the role of women and racial minorities both on the air and in the managerial ranks. Talk shows that would not have existed without the citizens push for access now have become staples of local television and the means for ventilating a wide range of views that previously had no outlets. Virtually every station in the country now has a daily or weekly talk program for local issues, under such titles as *Scope, Panorama,* or the name of the host (*The Stanley Siegel Show*), and some have programs devoted to black affairs. Such nationally syndicated or public-television series as *Not for Women Only, Black Journal,* William F. Buckley's *Firing Line, Black Perspective on the News* and *Women Alive* all owe their existence to the climate created by the citizens movement. Many stations have expanded the length of their local newscasts to carry segments on community issues, and increasingly they are featuring female and black reporters.

It is most noteworthy that public participation and the integration of the industry with minorities and women have not resulted in any economic hardship to broadcasting. Even with the loss of cigarette advertising and the reduction in children's commercials, television and radio have been setting new records for revenues and profits year after year.

The simple, crucial fact about American broadcasting is that it cannot be reformed from within. The industry perceives television as a great *business,* not as a great communications medium. In the real world, the corporate powers do not on their own elect to settle for 6 percent less profit one year in order to do better by the people. Any chairperson or president who would make such a decision would court serious trouble with the shareholders and could be removed from office. Because individual careers are always on the line, the business is geared for short-range survival decisions instead of long-range thinking that might make for a better medium and a healthier industry.

In short, broadcasting in America is lashed to a system the practitioners could not reform even if they wanted to. Whatever changes occur in the public interest will be brought on by forces outside the industry—consumer groups, minority coalitions, public-interest law firms, and organizations of concerned viewers—provided they have done their homework, made judgments that have moral force, brought action, and persevered.

Sooner or later the indicated reforms will be implemented by the FCC, the FTC, Congress, or the courts. Broadcasters will react by bellowing about the ruination of their industry—as they did when cigarette advertising was legislated off the airwaves in the public interest, or when children's advertising was curtailed—but they will quickly learn to live with the changes, and they will still prosper.

The future character of American broadcasting will be shaped by the growing tension between the industry and its public. Out of that most desirable tension should come a tensile strength that in time could transform mediocrity into greatness.